SpringerBriefs in Materials

The SpringerBriefs Series in Materials presents highly relevant, concise monographs on a wide range of topics covering fundamental advances and new applications in the field. Areas of interest include topical information on innovative, structural and functional materials and composites as well as fundamental principles, physical properties, materials theory and design. SpringerBriefs present succinct summaries of cutting-edge research and practical applications across a wide spectrum of fields. Featuring compact volumes of 50 to 125 pages, the series covers a range of content from professional to academic. Typical topics might include:

- A timely report of state-of-the-art analytical techniques
- A bridge between new research results, as published in journal articles, and a contextual literature review
- A snapshot of a hot or emerging topic
- An in-depth case study or clinical example
- A presentation of core concepts that students must understand in order to make independent contributions

Briefs are characterized by fast, global electronic dissemination, standard publishing contracts, standardized manuscript preparation and formatting guidelines, and expedited production schedules.

More information about this series at http://www.springer.com/series/10111

Angel Yanguas-Gil

Growth and Transport in Nanostructured Materials

Reactive Transport in PVD, CVD, and ALD

Springer

Angel Yanguas-Gil
Northwestern-Argonne Institute of Science
and Engineering
Northwestern University
Evanston, IL
USA

ISSN 2192-1091 ISSN 2192-1105 (electronic)
SpringerBriefs in Materials
ISBN 978-3-319-24670-3 ISBN 978-3-319-24672-7 (eBook)
DOI 10.1007/978-3-319-24672-7

Library of Congress Control Number: 2016953644

Printed on acid-free paper

This Springer imprint is published by Springer Nature
The registered company is Springer International Publishing AG
The registered company address is: Gewerbestrasse 11, 6330 Cham, Switzerland

A mis chicas: Elsa, Carmen y Ana

Preface

The central topic of this book is the coating and modification of high surface area materials using vapor phase deposition methods. This ability is a fundamental component of the nanomanufacturing toolbox, both at a research level, for instance to design of architectures and catalyst materials, and from a manufacturing point of view, as an integral component of semiconductor processing. It is also a field that has a long history that extends back to the 1960s in the early years of semiconductor processing, and even earlier, to the beginning of the twentieth century, if we include the fundamentals of heterogenous catalysis and rarefied gas flow. This diversity of applications and rich history makes it sometimes difficult for researchers in one area to take advantage of advances made in a different field: as much as research is becoming more interdisciplinary, a cursory look at the literature shows how discoveries and models have been re-discovered in different fields, or how problems that were once solved in a particular domain pop up later in a different place.

The purpose of this book is to bring together the fundamentals of the coating and functionalization of high surface area and nanostructure materials cutting across all these disciplines and application domains. The hope is that this book will be helpful to those interested in applying these techniques to fabricate new architectures or in developing new processes that can enable new materials and devices. I don't expect this book to be the definitive answer to this problem: due to the breadth of the subject and the limitations in space, it is impossible to explore in depth all of the topics included in this work, many of which would deserve their own books. Others didn't even make it into the manuscript. The same goes with the references included in this book: many outstanding examples in the literature had to be left out, including some of my personal favorites.

I have divided the material into five different chapters: Chap. 1 provides an introduction to the problem, including the range of substrates, growth techniques, and applications. Chapter 2 introduces the main deposition techniques, with a particular emphasis on those aspects that are relevant to the growth inside nanostructured or high surface area materials. Chapter 3 introduces the fundamentals of gas transport inside nanostructured materials: ballistic transport, diffusive models, and the impact of surface adsorption. Chapter 4 focuses on the application of these models to understand the growth inside high surface area materials. In particular, it emphasizes the importance of Thiele modulus as the key parameter controlling the infiltration dynamics in both chemical vapor deposition and atomic layer deposition. The chapter also presents a criterion to achieve conformal deposition for a given surface kinetics. Finally, in Chap. 5 I introduce two separate aspects of the coating of high surface area materials: the first one are modeling approaches to shape and surface evolution in high surface area and nanostructured materials. The second one is the coupling between the pore size scale and the reactor scale to understand the impact that high surface area materials can have in the transport of materials inside thin film growth reactors.

The main motivation for writing this book has been my own research experience: moving across application domains and growth techniques in the area of thin films and nanomaterials has allowed me to explore the fundamentals of reactive transport inside nanostructured materials from different perspectives, and work with processes with vastly different surface kinetics. I also perceived a gap in the literature, a single point of reference for the topics treated in this book. My approach to this book and the choice of topics have also been strongly influenced by my training as a theoretical physicist: when it comes to developing new processes or transferring technology from research laboratories into manufacturing, I firmly believe that a better grasp of the fundamentals can help us understand how to best design or scale up a particular process.

Finally, I would like to acknowledge the Northwestern Argonne Institute of Science and Engineering for providing me with the right framework to write this book, a task that falls outside the scope of my work as a researcher at Argonne National Laboratory.

Northbrook, IL, USA
June 2016

Angel Yanguas-Gil

Contents

Chapter 1
Introduction

In the last twenty years the coating and modification of non-planar substrates using vapor phase deposition techniques has become a key process for the design and fabrication of novel materials and architectures. This chapter provides an overview of these methods and their range of applications. Section 1.2 introduces the main thin film growth techniques and the main fundamental concepts involving the reactive transport in the vapor phase, and a brief introduction to the types of substrates is provided in Sect. 1.3. Finally, Sect. 1.4 provides a historical overview.

1.1 Introduction

The ability to evenly coat high surface area and nanostructured materials is an enabling capability that has had a great impact across many application domains. Semiconductor processing, membrane functionalization, microelectromechanical systems (MEMS), the design and stabilization of nanostructured electrodes for photovoltaics and energy storage, the development of novel catalysts, and development of sensors are some of the areas of applications [1–6].

While there are many ways in which the ability to coat high surface area materials has been leveraged to design and fabricate new architectures, we can classify these examples into four different categories, summarized in Fig. 1.1:

- It can be used to *functionalize* the surface of high surface area materials, for instance changing its acidity or decorating the surface with metal nanoparticles. Another example is the passivation or stabilization of surfaces, such as the case of battery electrode materials or to reduce the sintering of nanoparticles [7]. The modification of porous materials is also important in catalysis [8].

A. Yanguas-Gil, *Growth and Transport in Nanostructured Materials*,
SpringerBriefs in Materials, DOI 10.1007/978-3-319-24672-7_1

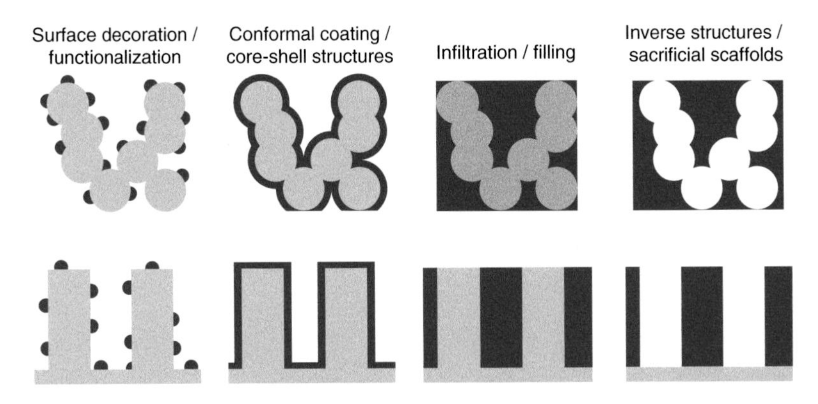

Fig. 1.1 Main applications of the coating of high surface area materials

- It can be used to create a *multilayer stack* over high surface area materials. Examples of these can be found in barrier and liner coatings for high aspect ratio trenches or vias, the design of cores-shell nanostructured electrodes, or the formation of 3D heterojunctions [9].
- It can be used to completely *infiltrate and fill* the voids of a 3D structure. Examples include the creation of nanocomposite materials or gap filling in semiconductor industry [10, 11].
- In combination with patterning and substractive techniques, it can be used to create nanostructured materials in which the original 3D or nanostructure is used as a *sacrificial scaffold*. Examples include the creation of inverse opals for photonic applications, or the synthesis of nanowires [12, 13].

As an example of the kind of designs that can be achieved, let us consider the core-shell structure shown in Fig. 1.2: by coating the high surface area core with a shell of a different material, we can impart a new functionality to its surface. Through such a design we can decouple the structure of the material from its surface properties, create complex 3D interfaces, or break down the functionality of the material into two or more components. This structure also presents a clear advantage with respect to the alternative approach of coating individual particles: the core shell structure in Fig. 1.2 does not break the continuity of the core, something which can have a great impact for instance in applications where charged species must be transported through the core.

Regardless of the application considered, the fundamental problem underlying the fabrication of these structures is the same: the growth inside high surface area materials (or any 3D substrate for that matter) relies on a delicate balance between the ability to efficiently transport gaseous species inside high surface area materials and the reactivity of the molecules with both with the inner surfaces and in the gas phase. From an atomistic point of view, when species from the gas phase penetrate the pores of a nanostructured material, they undergo multiple collisions with the walls until they finally reach the surface. From this perspective, the different chemical and

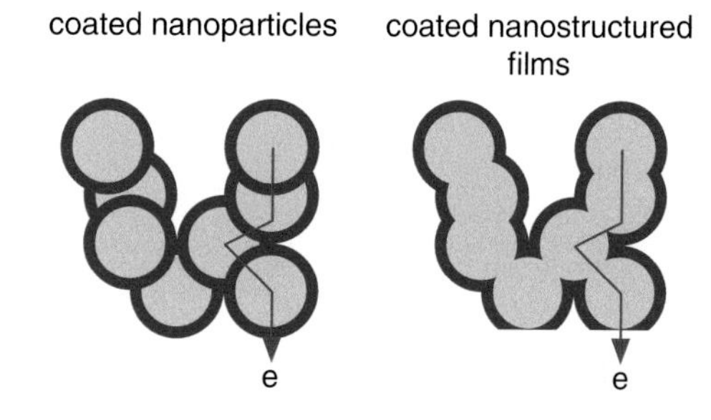

Fig. 1.2 Coating nanostructured electrodes allow the formation of a continuum path for electric carriers, something that is not possible when particles are coated individually

physical vapor deposition methods are just different approaches by which we can generate and transport reactive species to the substrate.

1.2 Vapor Phase Thin Film Deposition Techniques

The focus of this book is on vapor phase thin film growth techniques. These methods include techniques such as evaporation, sputtering, chemical vapor deposition (CVD), chemical vapor infiltration (CVI), and atomic layer deposition (ALD). With the exception of chemical vapor infiltration, which is a variant of CVD used in the synthesis of composite ceramic materials, these techniques play an important role in semiconductor manufacturing and are present in most of nanofabrication labs. These techniques are explored in more details in Chap. 2.

While all these approaches can be viewed as different means to achieve the efficient transport of materials inside nanostructured or high surface area material, there is an important conceptual difference between atomic layer deposition and the other techniques: even though pulsed processes have been developed for all of them, evaporation, sputtering, CVD, and CVI are essentially steady state techniques: when left alone, growth proceeds at a constant growth rate as a function of time. Atomic layer deposition, on the other hand, is based on the self-limited interaction of species with the surface: molecules can only react with a finite number of surface sites, and once these are consumed, the growth stops. By alternating exposures of two or more different gases, it is possible to regenerate the surface and start this cycle all over again. This causes atomic layer deposition to have certain unique properties, including its ability to be intrinsically conformal, and being highly scalable.

1.2.1 Conformality of Vapor Phase Deposition Techniques

From a quantitative perspective, the conformality of a process can be defined in terms of the variation on thickness or growth rates at different points of the material.

Table 1.1 Comparison between different vapor phase deposition methods

Process	Conformality	Aspect ratio	Growth rate	Comments
Evaporation	Poor	1:1	Fast	Line of sight
Sputtering	Low	1:5	Fast	Best results are obtained with iPVD
CVD	Depends	1:50	Medium	Extremely process-dependent
ALD	Excellent	1:1000	Slow	Well-behaved processes intrinsically conformal

The concept of *step coverage* can be formally defined as the ratio between the thickness of the film at the bottom and at the top of a feature or high surface area material:

$$\mathrm{SC} = \frac{\text{thickness at the bottom}}{\text{thickness at the top}} \tag{1.1}$$

A step coverage SC $= 1$ implies a perfectly conformal process.

Another relevant parameter is the term *fill factor*, defined as the ratio of the thickness of the film at the bottom of the feature to the nominal thickness of the film plus the depth of the via. This parameter was introduced in the context of semiconductor processing and it is a measure of the degree of planarization that can be achieved when growth is promoted at the bottom of the trench.

Roughly speaking, the ability to evenly coat high surface area material increases from evaporation (very poor), to atomic layer deposition (intrinsically conformal in well-behaved cases). However, as shown in Chap. 2, each technique provides its own way of optimizing and tailoring processes to improve their ability to coat high surface area materials.

Table 1.1 summarizes the capabilities of each of the techniques mentioned above. The values provided represent an order of magnitude of what a good, optimized process can achieve in each case. It is by no means an absolute maximum. Likewise, the conformality of a process can be much poorer if the process is run in suboptimal conditions.

In the case of PVD, early works showed the challenge of uniformly coating steps using evaporation methods [14]. At the same time, Hamaguchi and Rossnagel demonstrated the ability to conformally coat trenches of aspect ratio 4 with various metals using ionized PVD [15].

Although CVD is not a strictly conformal deposition technique, it has successfully being applied to fabricate inverse opal structure, such as Ge using digermane or HfB_2 using $Hf(BH_4)_4$ [12, 16]. It is also possible to take non-conformal CVD processes and improve their conformality by tweaking their surface kinetics. For instance, Kumar et al. showed how the conformality of a CVD process for the deposition TiB_2 could be improved by tailoring its sticking probability [17].

Finally, even in the case of ALD there are cases in which the conformality of an ideal ALD process is hard to achieve: while Elam et al. infiltrated carbon aerogels with surface area greater than 200m^2/g with tungsten using ALD [18], plasma assisted ALD faces challenge to achieve conformality in high aspect-ratio features when the recombination probability of plasma species is high [19].

1.2.2 *Sticking Probability*

A concept that is used to codify the reactivity of any given species with a surface and that will be used through this book is the *sticking probability*, defined as the probability that an incoming molecule reacts with the surface. Let ϕ be the incident flux per unit area of gas phase species to the surface, and ϕ_r the flux per unit area *coming out* from that surface. In the case of a non-reacting surface, we have that $\phi = \phi_r$. However, when walls are reactive, the outgoing flux will be only a fraction of the incoming flux. Therefore, we can define the *sticking probability* β, so that

$$\phi_r = (1 - \beta)\phi \tag{1.2}$$

Consequently, when $\beta = 1$, all the incoming particles react with the surface. When $\beta = 0$, the gas is non-reactive.

It is important to emphasize that behind the concept of sticking probability there are two different mechanistic interpretations: the sticking probability can be the consequence of a single particle-surface interaction, but it can also be defined in an statistical sense, as the balance between the incident and outgoing fluxes of a given species. In practice it is very hard to discriminate between two cases, since some species can have long residence times at the surface and undergo other kinetic processes before desorbing from the surface. In Chap. 4, we focus on the impact that the sticking probability has on the conformality of both non-self limited and self-limited processes (Fig. 1.3).

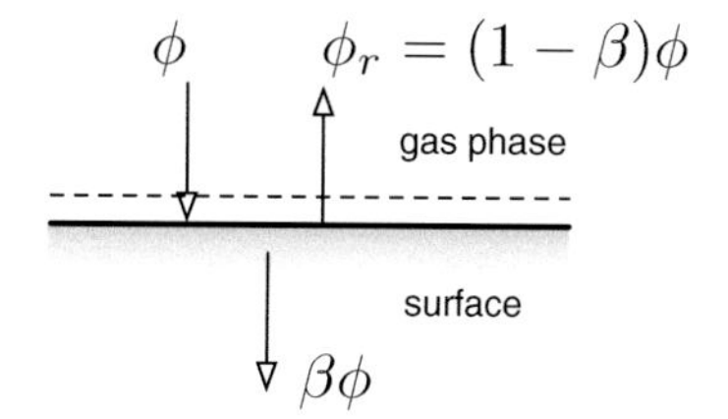

Fig. 1.3 The sticking probability β is a direct measure of precursor-surface interaction, playing a key role in the ability of a process to coat high surface area materials

1.2.3 Knudsen Number

A second important parameter to understand the reactive transport inside high surface area materials that takes place in vapor phase thin film growth is the *Knudsen number*, defined as the ratio of the mean free path λ to the characteristic size of the pore d:

$$\mathrm{Kn} = \frac{\lambda}{d} \tag{1.3}$$

The mean free path in a homogeneous Maxwellian gas is given by:

$$\lambda = \frac{k_B T}{\sqrt{2}\sigma_c} \tag{1.4}$$

where σ_c is the collision cross section, k_B is Boltzmann's constant, and T is the gas temperature.

Figure 1.4 shows the boundary corresponding to $\mathrm{Kn} = 1$ as a function of characteristic size d for the case of nitrogen. For the bulk of applications considered in the literature the Knudsen number is of the order or greater than one, which means that the flow takes place in the rarified or Knudsen flow regime. However, this is not the case for atmospheric pressure processes, for which the flow is still collisional when feature sizes are of the order of 1 micron.

For Knudsen numbers greater than one, the gas phase collisions between molecules can be ignored inside a feature. The transport it is said to take place in the *Knudsen regime* or under *molecular flow*. As shown in Chap. 3, the reactive transport

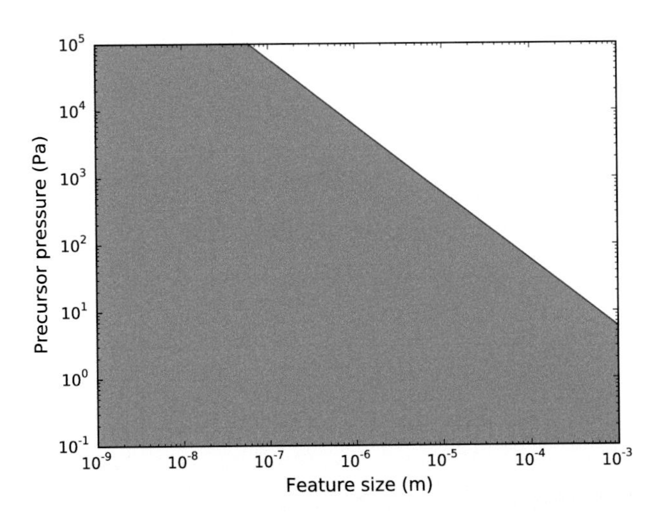

Fig. 1.4 Threshold pressures for Knudsen flow as a function of characteristic feature size for the case of nitrogen gas at room temperature. The *shaded area* corresponds to the Knudsen regime, $\mathrm{Kn} > 1$

of molecules inside the high surface area material becomes independent of the actual dimensions of the pores or features in the substrate. From a fundamental perspective, the transport can be described using ballistic models, and the same approaches and tools that have been developed to model rarified gas dynamics can be applied at the nanoscale.

If we consider the case of transport within a single cylindrical pore, as shown in Chap. 3 it is possible to define a diffusion coefficient that extends over the whole pressure range from the Knudsen to the viscous flow regimes using the Bosanquet approximation. In Fig. 1.5, we show the diffusion coefficient of TMA as a function of pore diameter at room temperature for three nitrogen background pressures: 1, 10, and 760 Torr. This shows how as the flow transitions from Knudsen to molecular, the diffusion coefficient becomes independent of the pore size. The transition takes place at diameters satisfying the condition $\mathrm{Kn} = 1$.

In addition to the transport inside the pore, the mean free path plays an important role in determining the distribution of incident species. In techniques such as evaporation or molecular beam epitaxy the pressure is so low that the sources used for growth have a strong directional component. Such directional component is sometimes beneficial, for instance to reduce sidewall deposition in trenches and vias. In contrast, when the gas can be considered in local thermodynamic equilibrium, the flux of incident species to the surface can be approximated as:

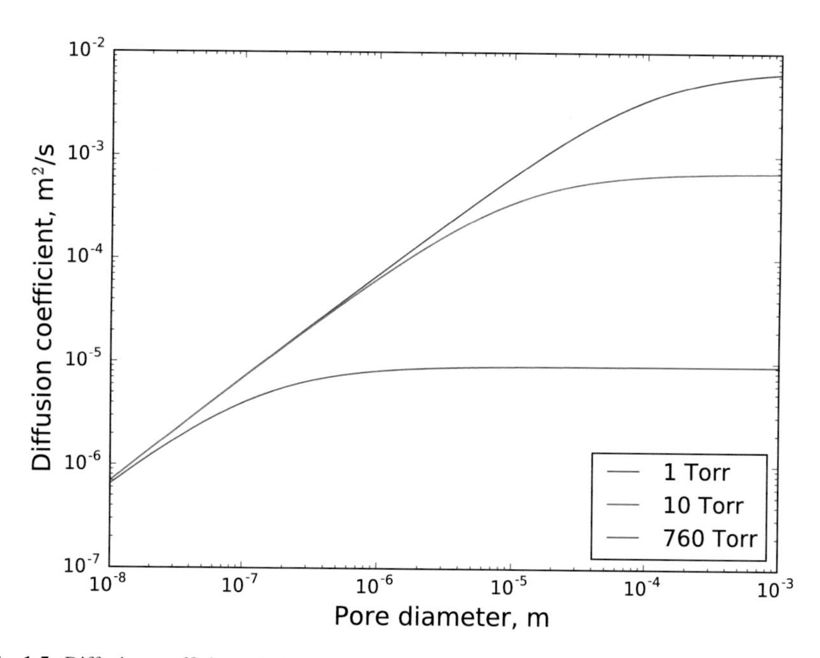

Fig. 1.5 Diffusion coefficient of trimethylaluminum inside a pore as a function of pore diameter for different nitrogen background pressures: as molecular diffusion dominates, the diffusion coefficients becomes independent of pore size

$$\phi = \frac{p}{\sqrt{2\pi M k_B T}} \tag{1.5}$$

where p is the pressure of the gas phase species, T is the temperature, and M the molecular mass. Controlling the incident distribution function of species is one of the ways in which processes that are essentially line of sight can be adapted to the coating of structured substrates.

1.3 Substrates and Scaffolds

There is a wide range of high surface area and nanostructured materials that have been used in the literature as substrates. In the context of this book, it is useful to break them into two categories: deterministic substrates with a well-defined structure, and nanostructured or high surface area materials with a random morphology (Fig. 1.6).

In both cases, the atomistic aspects of reactive transport taking place during thin film growth are essentially the same. However, in geometrical substrates the evolution of the surface during growth can be accurately modeled if the local growth rate at every point of the surface is known. In contrast, in a material with a random porosity or morphology we need to resort to other approaches to take into account the random nature of the structure of the material. A challenge of non-deterministic substrates is how to determine the evolution of the inner surface as the thickness of the materials increases.

A common concept to both of them is the *surface area enhancement* of a substrate s_e, defined as the surface area per unit (projected) surface area. This concept is related to the so-called *specific surface area*, defined as the surface area per unit volume $\bar{s}$ in the material, through the equation:

$$s_e = \bar{s}L \tag{1.6}$$

where L is the thickness of the material. Note that there are two common definitions of specific surface area: one defined in terms of volume, and a second one defined in terms of mass. The two are related through the density of the material. In the models

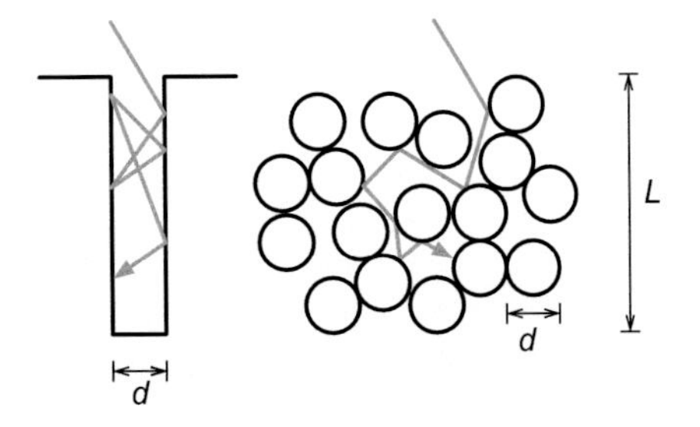

Fig. 1.6 Two types of two surface area materials: geometrical features and porous materials with random microstructure

introduced in Chaps. 3 and 4, the specific area $\bar{s}$ defined in terms of the volume of the material is the relevant parameter.

1.3.1 *Deterministic Substrates*

Trenches and vias are by far the best studied systems in the literature, due to their central role in semiconductor manufacturing [20–22]. Other examples in the literature of the coating of trenches and circular vias include the epitaxy of III-V semiconductors on trenched substrates [23, 24], the fundamental growth studies and model devices fabricated on anodised alumina membranes [25, 26], and capillary glass arrays functionalized to fabricate microchannel plates for photodetectors [27]. Anodised aluminum oxide has also been used as a template for nanowire array fabrication using pore infiltration [28] (Fig. 1.7).

For these substrates, the key variable is the *aspect ratio*. In the case of a rectangular trench, the aspect ratio is defined as the depth to width ratio, while for circular pores it is defined as the depth to diameter ratio:

$$\mathrm{AR} = \frac{L}{d} \tag{1.7}$$

The reason why this parameter is so relevant is that, as described in Chap. 3, at low enough pressures the reactive transport process becomes independent of the actual size of a feature.

The aspect ratio is also a measure of the surface area enhancement of a feature. In the case of a rectangular trench, the cross sectional area is given by $S_0 = dl$, where l is the length of the trench. If we assume that the trench is much longer than wide,

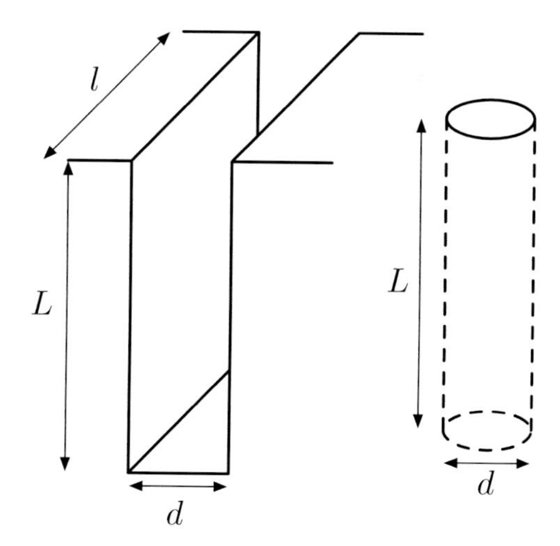

Fig. 1.7 Key geometric parameters in *rectangular trenches* and *circular pores* or vias

$l \gg d$, then the total surface area of the trench is $S_t \approx 2lL + S_0$. Consequently, the surface area enhancement s_e is given by:

$$s_e = \frac{S_t}{S_0} = 1 + \frac{2L}{d} = 1 + 2\text{AR} \tag{1.8}$$

Likewise, for a circular pore or via, $S_v = \pi dL + S_0$, where d is the diameter of the pore, so that:

$$s_e = \frac{S_v}{S_0} = 1 + \frac{4L}{d} = 1 + 4\text{AR} \tag{1.9}$$

Note that these values are for a single pore. This surface enhancement values need to be modulated by the relative surface area occupied by the high surface area structures on the substrate [29].

Deterministic substrates go well beyond pores and vias. In particular, in the area of colloidal crystals, inverse opals and woodpile structures have been synthesized using vapor phase infiltration techniques, both chemical vapor deposition and atomic layer deposition. Other structures include periodic nanowire and nanorod arrays [30–32].

The formation of some of these structures relies on the ability to infiltrate sacrificial scaffolds. This is the case of tungsten photonic band gap structures synthesized using tungsten CVD followed by etching [33]. Inverse opals have also been realized by infiltration of sacrificial opals both by CVD and ALD [12, 34].

Opals are formed by spheres arranged in a closed pack structure. Therefore, the surface area enhancement of an opal film will depend only on its number of layers N through the formula:

$$s_e = \frac{4\pi N}{3\sqrt{2}} \approx 2.96 \times N \tag{1.10}$$

More recently, rectangular and triangular fins have become central to novel three dimensional field effect transistors such as the FinFET, shown in Fig. 1.8 [35]. In particular, the fabrication of the metal gate stack requires the conformal coating of the fin with a gate dielectric layer (colored in yellow in Fig. 1.8).

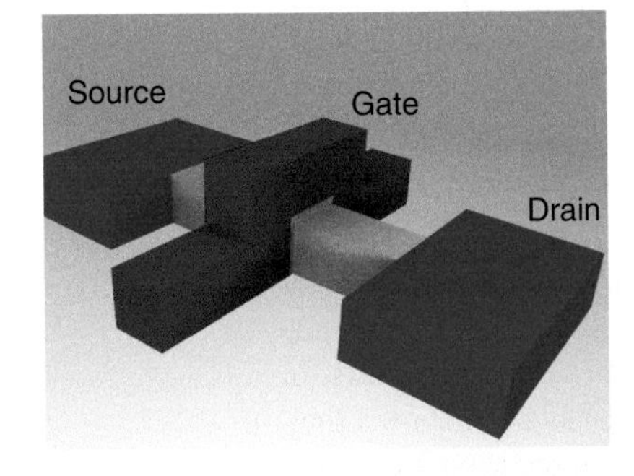

Fig. 1.8 Scheme of a finFET: both the fabrication of the fin and the fabrication of the gate dielectric (*yellow layer* in the *diagram*) involves the conformal deposition over non-planar surfaces

Finally, 3D ordered mesoporous solids such as zeolites and metalorganic frameworks (MOFs) represent an extreme case of ordered material in which the pore diameters are comparable to the diameter of a precursor molecule. Zeolites have been used as templates for the synthesis of nanoporous materials using chemical vapor deposition [36–38]. Examples of the use of metallorganic frameworks (MOFs) as substrates include the metalation with trimethylaluminum of single crystal NU-1000 MOFs [39].

1.3.2 Porous and Disordered Substrates

The second class of high surface area materials are those exhibiting a random microstructure. This includes the whole spectrum of porous materials but also nanowire and nanotube forests, both vertically and randomly aligned.

When describing porous materials, in the context of this book it is important to distinguish between total and effective porosity: while total porosity refers to the total volume fraction occupied by void, *effective porosity* ε is defined as the volume of interconnected pores that are accessible to gas molecules. This is the relevant parameter in the context of thin film growth.

The *pore size distribution* of a porous material provides the volume of pores with a characteristic size between d and $d + \Delta d$. The pore width d is defined as either the diameter of a cylindrical pore, or the distance between opposite walls as in the case of slit-shaped pores. Depending on its size, IUPAC defines three types of pores: *micropores* (0.3–2 nm), *mesopores* (2–50 nm) and *macropores* ($>$50 nm).

Particle beds are one of the most common examples of random substrates. The conformal coating of fixed particle beds allows for the modification of particles at the lab scale without the need of using more complex experimental setups for particle coatings involving agitation or fluidization. Most of the examples in the literature have focused on atomic layer deposition, and two key areas of applications are the stabilization of cathode and anode materials for lithium ion batteries, and the design of new catalysts, in which ALD is used both for the catalyst support and to synthesize metal nanoparticles [40, 41]. The latter takes advantage of the nucleation behavior of some metal ALD processes to form nanoparticles instead of fully continuous films.

Nanostructured and mesoporous films play a central role in areas such as energy storage or in nanostructured photovoltaics. One well-known example is the nanostructure electrode in dye sensitized solar cells, typically based on sintered titania anatase nanoparticles with a characteristic size of 20 nm. The coating of nanostructured electrodes allow for the design of core shell structures such as those shown in Figs. 1.1 and 1.2 [42, 43]. The key advantage of coating pre-defined mesoporous structures rather than individual particles is that it allows the formation of a continuous path for carriers. In fact, in the stabilization of lithium-ion batteries using Al_2O_3 it has been shown that the coating of electrodes provides superior properties when compared to individually coated particles [7].

Another example are materials based on 1D constituents, such as nanowire arrays, nanotubes or fiber bundles. Semiconductor nanowires have been extensively used. Examples include the hematite coating of Si nanowires by atomic layer deposition [44]. Likewise, sulfide-nanotube composite materials synthesized by ALD have been used as anode materials for energy storage applications [45].

Low temperature processes enabled by ALD have allowed the coating and modification of fiber-based organic materials, including textiles [46, 47]. Finally, fiber bundles are typically used as starting materials for the formation of nanocomposite ceramic materials using chemical vapor infiltration [10].

While not traditionally considered porous materials, polymers represent the extreme case of a porous disordered material. The permeability of polymers to different gases has long been the subject of active research and industrial interest. Infiltration approaches in the literature have taken advantage of this permeability to fabricate hybrid organic/inorganic materials by pairing polymers with precursors that are reactive towards the functional groups within the polymers [48]. One application that has taken advantage of the permeability of polymers is the fabrication of photoresists for semiconductor manufacturing based on the infiltration of block copolymers [49, 50].

1.4 Historical Overview

The fundamentals of reactive transport in nanostructured materials is a topic that cuts across multiple disciplines that in some cases do not overlap. This issue was already emphasized more than thirty years ago by Steckelmacher [51], who reviewed some of the fundamental aspects of Knudsen flow, and cited rarefied gas dynamics, vacuum and surface science, and thin film deposition as some of the relevant areas.

While the origin of molecular flow can be traced back to the works of Knudsen in 1909, here we are going to emphasize three areas in which the main concepts for understanding the reactive transport and growth inside nanostructured materials were developed: semiconductor manufacturing, chemical vapor infiltration, and chemical engineering and catalysis.

1.4.1 Conformality in Semiconductor Manufacturing

The fabrication of integrated circuits has long required the ability to deposit homogeneous films over various topographic features. While early patterning strategies based on liftoff techniques took advantage of line of sight deposition process to expose the edge of the resist to the dissolving chemicals, the formation of metal contacts with the underlying silicon required the deposition of a metal film, typically aluminum, over a dielectric step (Fig. 1.10). Thickness inhomogeneity would lead to either to an open contact or to device failure caused by electromigration, an early reliability concern [52]. This is the origin of the term *step coverage*, originally applied in its literal sense to describe the ability of a process to coat homogeneously a step on a surface (Fig. 1.9).

Metallization was one of the areas with a strong need of processes able to coat, and in some cases infiltrate, high aspect ratio features. If we use the tungsten plug process as an example, CVD was the preferred approach to establish the contacts with the heavily doped areas of silicon. In addition to the tungsten deposition step itself, the W plug process required the deposition of other layers to engineer the contact and to protect the underlying material from the WF_6 precursor [53]. Some of these layers were still fabricated using PVD methods, such as sputtered Ti, requiring the development of novel strategies to improve the conformality of sputtering on high aspect ratio trenches and vias.

Likewise, the introduction of low-k and porous silicon together with the transition from Al to Cu required depositing liners and diffusion barriers inside trenches and vias to prevent Cu diffusion into the dielectric [9]. This lead to the development of numerous CVD and ALD processes for barrier materials, for which a good conformality was one of the key requirements. It also motivated a lot of research on Cu CVD and ALD as a way of depositing seeds for the subsequent trench and via filling using electrodeposition.

Another technology driver was the introduction of the trench and stack capacitor concepts for DRAMs. Each DRAM cell is composed of a MOS transistor and a storage capacitor. While originally planar capacitors were used in the early designs, in 1978 Koyanagi et al. first proposed the stacked capacitor, which came into production around 1985 [54]. Since then the increase in DRAM density has lead to both an increase in aspect ratios, and the incorporation of new electrode materials and dielectrics, with poly-Si, SiO_2 and later Ta_2O_5 being replaced by high-k dielectrics and metals such as TiN as the design moved from PIP to MIM capacitors. This evolution has driven the development of numerous processes for dielectrics and metals using CVD and ALD. More recently, 3D architectures such as those used in vertical flash require the growth of highly conformal films.

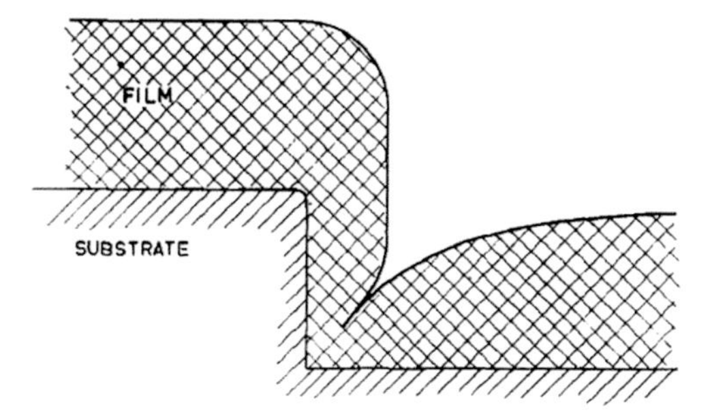

Fig. 1.9 Simulated profile of an evaporated film on a step. This is the origin of the term *step coverage* often used to describe the ability of a process to homogeneously coat high aspect ratio or high surface area materials. Reproduced from Blech with permission [14]

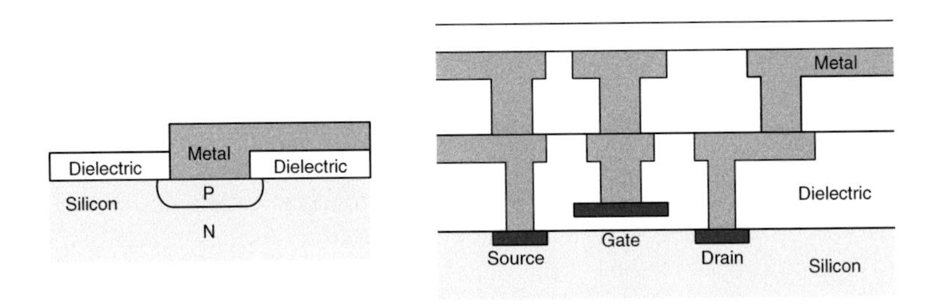

Fig. 1.10 Schemes of cross-sectional overviews of a metal-silicon contact and a multilever metallization scheme

A final third area of interest in semiconductor processing was the development of shallow trench insulation (STI) for device insulation. The increasing scale down of device dimensions spurred the search of new approaches to insulate devices that allowed a more efficient packing than the traditional LOCOS method. The shallow trench insulation approach required dielectric gap fill of trenches of moderate aspect ratio trenches patterned on the Si wafer, with SiO_2 the most common choice. The ability to fill these trenches was demonstrated for instance using tetraethoxysilane and ozone both at subatmospheric and low vapor pressure deposition. However, high density plasma CVD methods capable of complete gap filling have also been developed due to the higher density of the deposited films, which avoid the need of a densification step [11, 21].

1.4.2 Chemical Vapor Infiltration

Chemical vapor infiltration (CVI) is a technique used for the synthesis of novel ceramic materials. It applies chemical vapor deposition to the infiltration of high surface area materials to form composite materials. The goal of this process is to enhance the density of the material, typically by a factor of 2 to up to 10, starting from a porous preform such as a bundle or array of fibers. This technique has been applied to the synthesis of composites for aircraft brakes and other aerospace applications [55].

Isothermal, isobaric CVI was one of the earliest approaches, being developed in the 1960s. The range of materials deposited include C, SiC, TiC, Si_3N_4, BN, Al_2O_3, and ZrO_2, though the bulk of research has been carried out for C and SiC [56]. Improving the infiltration dynamics and increasing the throughput of the overall process have always been concerns in CVI, and several refinements have been proposed to improve the efficiency of the process. One example are forced-flow methods, in which pressure and temperature gradients are used to increase the homogeneity of the infiltration process [10] (Fig. 1.11).

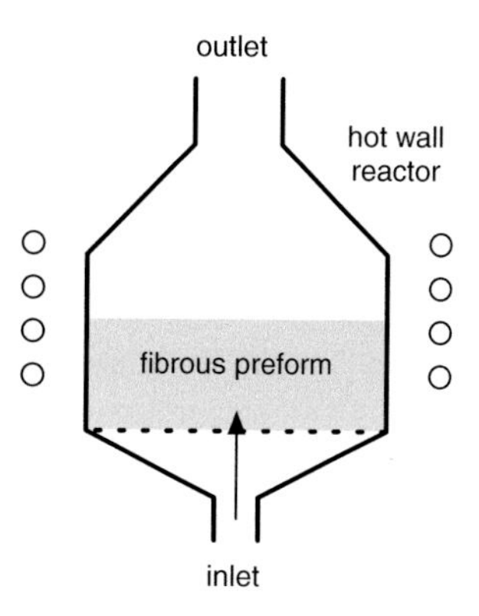

Fig. 1.11 Experimental setup of a typical CVI reactor

As part of the development of this technique, the fundamentals of infiltration under CVD conditions were studied both in presence and in absence of homogeneous reactions, establishing a direct connection with the approaches to model reactive transport in heterogeneous catalysis [57].

1.4.3 Chemical Engineering and Heterogeneous Catalysis

One of the key contributions of chemical engineering and heterogeneous catalysis to the topic of this book was the development of the foundations of both no-reactive and reactive transport in high surface area materials, with the seminal contributions of Damköler and Thiele going back to the 1930s [58].

Many of the efforts in this field have focused on developing good diffusivity models that would reflect the underlying microstructure of porous materials. Catalysts and catalyst supports in heterogeneous catalysis can take a wide range of forms, including porous pellets, engineered monolyths, and zeolites, each with its own particular microstructure. Therefore, a lot of effort was devoted to understanding the transport and the reaction of gases in porous materials.

In particular, a key concept that, as shown in Chap. 4, will also play a role in the coating of nanostructured materials is the *Thiele modulus* h_T. It is defined as the ratio of intrinsic reaction rate to the diffusion rate. The definition for a cylindrical pore of length L and diameter d for a first order reaction rate k given by:

$$h_T = \sqrt{\frac{4kL^2}{dD}} = \sqrt{\frac{4(\mathrm{AR})L}{D}} \tag{1.11}$$

Where D is the diffusion coefficient of the precursor molecule inside the nanopore. Likewise, the *effectiveness or utilization factor* is defined as the ratio between the averaged effective reaction rate r_e divided by the intrinsic reaction rate r_0, which is obtained in the case in which the process is not affected by transport limitations:

$$\eta = \frac{r_e}{r_0} \tag{1.12}$$

In some cases the synthesis of the catalyst itself may require the infiltration of porous materials. For instance, as early as in 1940 the modification of silica gel by TiO_2 formed by the hydroxylation of $TiCl_4$ using a solution based process was reported in the literature, and in-situ infrared spectroscopy studies of the materials formed by the gas-phase infiltration of porous silica gel by trimethyl aluminum and other metal precursors and their subsequent hydroxylation were published in the late sixties and early seventies [59, 60]. CVD and ALD were eventually adopted as techniques for catalyst synthesis: catalyst synthesis approaches based on CVD started appearing in the literature in the 1980s. One of the earliest applications was the modification of silica gel with boron oxide using chemical vapor deposition [61]. Atomic layer deposition examples (at that time referred to as atomic layer epitaxy) appear in the scientific literature in the early 1990s [62].

1.5 Summary

The coating of high surface area material is a crosscutting capability that has found application in a wide range of application domains, often with little or no overlap between them. In this chapter we have provided an overview of the techniques and substrates used for the coating of high surface area and nanostructured materials, as well as the key concepts that are important to understand the fundamentals of reactive transport in nanostructured materials.

References

1. L. Niinisto, J. Paivasaari, J. Niinisto, M. Putkonen, M. Nieminen, Phys. Status Solidi A **201**(7), 1443 (2004)
2. M. Knez, K. Niesch, L. Niinisto, Adv. Mater. **19**(21), 3425 (2007)
3. F. Zaera, Chem. Soc. Rev. **42**(7), 2746 (2013)
4. A. Stein, B.E. Wilson, S.G. Rudisill, Chem. Soc. Rev. **42**(7), 2763 (2013)
5. C. Marichy, M. Bechelany, N. Pinna, Adv. Mater. **24**(8), 1017 (2012)
6. M.E. Donders, J.F.M. Oudenhoven, L. Baggetto, H.C.M. Knoops, M.C.M. van de Sanden, W.M.M. Kessels, P.H.L. Notten, in *Atomic Layer Deposition Applications 6, ECS Transactions*, vol. 33 (2010), pp. 213–222
7. Y.S. Jung, A.S. Cavanagh, L.A. Riley, S.H. Kang, A.C. Dillon, M.D. Groner, S.M. George, S.H. Lee, Adv. Mater. **22**, 2172 (2010)
8. V. Meille, Appl. Catal. A-Gen. **315**, 1 (2006)
9. A.E. Kaloyeros, E. Eisenbraun, Ann. Rev. Mater. Sci. **30**, 363 (2000)
10. D.P. Stinton, A.J. Caputo, R.A. Lowden, Am. Ceram. Soc. Bull. **65**(2), 347 (1986)
11. H. Nishimura, S. Takagi, M. Fujino, N. Nishi, Jpn. J. Appl. Phys. Part 1-Regul. Pap. Short Notes Rev. Pap. **41**, 2886 (2002)
12. H. Miguez, E. Chomski, F. Garcia-Santamaria, M. Ibisate, S. John, C. Lopez, F. Meseguer, J.P. Mondia, G.A. Ozin, O. Toader, H.M. van Driel, Adv. Mater. **13**, 1634 (2001)

13. K.B. Lee, S.M. Lee, J. Cheon, Adv. Mater. **13**(7), 517 (2001)
14. I.A. Blech, Thin Solid Film. **6**, 113 (1970)
15. S. Hamaguchi, S.M. Rossnagel, J. Vac. Sci. Technol. B **14**, 2603 (1996)
16. K.A. Arpin, M.D. Losego, A.N. Cloud, H.L. Ning, J. Mallek, N.P. Sergeant, L.X. Zhu, Z.F. Yu, B. Kalanyan, G.N. Parsons, G.S. Girolami, J.R. Abelson, S.H. Fan, P.V. Braun, Nat. Commun. **4**, 2630 (2013)
17. N. Kumar, A. Yanguas-Gil, S.R. Daly, G.S. Girolami, J.R. Abelson, J. Am. Chem. Soc. **130**(52), 17660 (2008)
18. J.W. Elam, J.A. Libera, M.J. Pellin, A.V. Zinovev, J.P. Greene, J.A. Nolen, Appl. Phys. Lett. **89**(5), 053124 (2006)
19. H.C.M. Knoops, E. Langereis, M.C.M. van de Sanden, W.M.M. Kessels, J. Electrochem. Soc. **157**(12), G241 (2010)
20. J.G. Ryan, R.M. Geffken, N.R. Poulin, J.R. Paraszczak, IBM J. Res. Dev. **39**(4), 371 (1995)
21. M. Nandakumar, A. Chatterjee, S. Sridhar, K. Joyner, M. Rodder, I.C. Chen, in *International Electron Devices Meeting 1998. Technical Digest* (Cat. No.98CH36217) (1998), pp. 133–136
22. S.P. Murarka, Mater. Sci. Eng. R-Rep. **19**(3–4), 87 (1997)
23. Y. Chen, R. Schneider, S.Y. Wang, R.S. Kern, C.H. Chen, C.P. Kuo, Appl. Phys. Lett. **75**(14), 2062 (1999)
24. C.I.H. Ashby, C.C. Mitchell, J. Han, N.A. Missert, P.P. Provencio, D.M. Follstaedt, G.M. Peake, L. Griego, Appl. Phys. Lett. **77**(20), 3233 (2000)
25. J.W. Elam, D. Routkevitch, P.P. Mardilovich, S.M. George, Chem. Mater. **15**(18), 3507 (2003)
26. A.B.F. Martinson, J.W. Elam, J. Liu, M.J. Pellin, T.J. Marks, J.T. Hupp, Nano Lett. **8**(9), 2862 (2008)
27. A.U. Mane, Q. Peng, J.W. Elam, D.C. Bennis, C.A. Craven, M.A. Detarando, J.R. Escolas, H.J. Frisch, S.J. Jokela, J. McPhate, M.J. Minot, O.H. Siegmund, J.M. Renaud, R.G. Wagner, M.J. Wetstein, Phys. Proc. **37**, 722 (2012)
28. D.J. Comstock, S.T. Christensen, J.W. Elam, M.J. Pellin, M.C. Hersam, Adv. Funct. Mater. **20**(18), 3099 (2010)
29. A.B.F. Martinson, J.W. Elam, J.T. Hupp, M.J. Pellin, Nano Lett. **7**(8), 2183 (2007)
30. D. Routkevitch, A.A. Tager, J. Haruyama, D. Almawlawi, M. Moskovits, J.M. Xu, IEEE Trans. Electron Dev. **43**(10), 1646 (1996)
31. H. Chik, J. Liang, S.G. Cloutier, N. Kouklin, J.M. Xu, Appl. Phys. Lett. **84**(17), 3376 (2004)
32. Y. Li, G.T. Duan, G.Q. Liu, W.P. Cai, Chem. Soc. Rev. **42**(8), 3614 (2013)
33. J.G. Fleming, S.Y. Lin, I. El-Kady, R. Biswas, K.M. Ho, Nature **417**(6884), 52 (2002)
34. A. Rugge, J.S. Becker, R.G. Gordon, S.H. Tolbert, Nano Lett. **3**, 1293 (2003)
35. D. Hisamoto, W.C. Lee, J. Kedzierski, H. Takeuchi, K. Asano, C. Kuo, E. Anderson, T.J. King, J. Bokor, C.M. Hu, IEEE Trans. Electron Dev. **47**(12), 2320 (2000)
36. Z.X. Yang, Y.D. Xia, R. Mokaya, J. Am. Chem. Soc. **129**(6), 1673 (2007)
37. H. Vuori, R.J. Silvennoinen, M. Lindblad, H. Osterholm, A.O.I. Krause, Catal. Lett. **131**(1–2), 7 (2009)
38. S. Hermes, M.K. Schroter, R. Schmid, L. Khodeir, M. Muhler, A. Tissler, R.W. Fischer, R.A. Fischer, Angew. Chem. Int. Ed. **44**(38), 6237 (2005)
39. J.E. Mondloch, W. Bury, D. Fairen-Jimenez, S. Kwon, E.J. DeMarco, M.H. Weston, A.A. Sarjeant, S.T. Nguyen, P.C. Stair, R.Q. Snurr, O.K. Farha, J.T. Hupp, J. Am. Chem. Soc. **135**(28), 10294 (2013)
40. L.A. Riley, S. Van Ana, A.S. Cavanagh, Y.F. Yan, S.M. George, P. Liu, A.C. Dillon, S.H. Lee, J. Power Sources **196**(6), 3317 (2011)
41. H. Feng, J.W. Elam, J.A. Libera, W. Setthapun, P.C. Stair, Chem. Mater. **22**, 3133 (2010)
42. B.M. Klahr, T.W. Hamann, J. Phys. Chem. C **113**(31), 14040 (2009)
43. Q.F. Zhang, C.S. Dandeneau, X.Y. Zhou, G.Z. Cao, Adv. Mater. **21**(41), 4087 (2009)
44. M.T. Mayer, C. Du, D.W. Wang, J. Am. Chem. Soc. **134**(30), 12406 (2012)
45. X.B. Meng, K. He, D. Su, X.F. Zhang, C.J. Sun, Y. Ren, H.H. Wang, W. Weng, L. Trahey, C.P. Canlas, J.W. Elam, Adv. Funct. Mater. **24**(34), 5435 (2014)

46. G.K. Hyde, K.J. Park, S.M. Stewart, J.P. Hinestroza, G.N. Parsons, Langmuir **23**(19), 9844 (2007)
47. J.S. Jur, W.J. Sweet, C.J. Oldham, G.N. Parsons, Adv. Funct. Mater. **21**(11), 1993 (2011)
48. G.N. Parsons, S.M. George, M. Knez, MRS Bull. **36**(11), 865 (2011)
49. Q. Peng, Y.C. Tseng, S.B. Darling, J.W. Elam, ACS Nano **5**(6), 4600 (2011)
50. Q. Peng, Y.C. Tseng, S.B. Darling, J.W. Elam, Adv. Mater. **22**(45), 5129 (2010)
51. W. Steckelmacher, Rep. Prog. Phys. **49**(10), 1083 (1986)
52. J.L. Vossen, G.L. Schnable, W. Kern, J. Vac. Sci. Technol. **11**(1), 60 (1974)
53. P.J. Ireland, Thin Solid Film. **304**, 1 (1997)
54. M. Koyanagi, H. Sunami, N. Hashimoto, M. Ashikawa, in *Electron Devices Meeting, 1978 International*, vol. 24, pp. 348–351 (1978)
55. I. Golecki, Mater. Sci. Eng. R-Rep. **20**(2), 37 (1997)
56. R. Naslain, F. Langlais, R. Fedou, J. Phys. **50**(C-5), 191 (1989)
57. S. Middleman, J. Mater. Res. **4**(6), 1515 (1989)
58. E.W. Thiele, Ind. Eng. Chem. **31**, 916 (1939)
59. R.J. Peglar, F.H. Hambleto, J.A. Hockey, J. Catal. **20**(3), 309 (1971)
60. D.J.C. Yates, G.W. Dembinsk, W.R. Kroll, J.J. Elliott, J. Phys. Chem. **73**(4), 911 (1969)
61. S. Sato, H. Sakurai, K. Urabe, Y. Izumi, Chem. Lett. **14**(3), 277 (1985)
62. S. Haukka, E.L. Lakomaa, O. Jylha, J. Vilhunen, S. Hornytzkyj, Langmuir **9**(12), 3497 (1993)

Chapter 2
Physical and Chemical Vapor Deposition Techniques

Gas phase deposition techniques are traditionally divided into two categories: physical vapor deposition methods, such as evaporation or sputtering, in which there is a net transfer of matter from a solid reservoir to a film, and chemical vapor deposition methods, where the film is formed as the product of the chemical reaction involving one or more species in the gas phase. The main techniques in the second category are chemical vapor deposition (CVD) and atomic layer deposition (ALD), both with thermal and plasma-assisted variations.

This chapter introduces these deposition techniques in increasing order of their ability to homogeneously coat high surface area materials: Sect. 2.1 focuses on physical vapor deposition (PVD) methods, Sect. 2.2 introduces chemical vapor deposition methods, while Sect. 2.3 focuses on atomic layer deposition. Since descriptions of these techniques can be found elsewhere, here we will highlight those aspects that are relevant for the coating of high surface area materials: the transport and reactivity of the species coming of the surface, and the strategies that have been developed to improve the ability to homogeneously coat high surface area materials. For details involving equipment and other aspects of the techniques the reader is referred to some of the monographs listed in the reference section.

2.1 Physical Vapor Deposition Methods: Evaporation and Sputtering

Evaporation and sputtering are two well-established manufacturing techniques whose origins can be traced back to the late 19th century. Both are considered to be mainly line-of-sight methods, meaning that they have a limited ability to coat high surface

A. Yanguas-Gil, *Growth and Transport in Nanostructured Materials*,
SpringerBriefs in Materials, DOI 10.1007/978-3-319-24672-7_2

area or high aspect ratio substrates as a consequence of their typically large sticking probabilities. Still, PVD methods have been successfully applied to coat moderate aspect ratio features, and have played a key role in areas such as semiconductor processing.

2.1.1 Evaporation

In vacuum evaporation, a net transfer of materials is created from a source, typically contained in a crucible, to the substrate by heating the crucible under high vacuum. Usually the pressure at which this process takes place is so low that the mean free path of the evaporated atoms is larger than the size of the chamber, resulting on a highly directional flux of species. This flux largely depends on the net evaporation rate, the angular distribution function of evaporated species, and the geometry of the deposition chamber.

For reasons that are explained more in detail in Chap. 3, a simple thermodynamic argument leads to an angular distribution of evaporated species that is proportional to the cosine of the angle normal to the surface. This relationship has been experimentally verified many times in the literature (Fig. 2.1). Consequently, if R is the distance between the crucible and the substrate, and θ_c and θ_s are the angles between the vector joining the crucible and the substrate and the surface normals of the crucible and the substrate, respectively, then the proportionality constant is given by:

$$k = \frac{\cos\theta_c \cos\theta_s}{\pi R^2} \tag{2.1}$$

This dependence affects the homogeneity of the growth across large areas. In particular, areas located directly on top of the crucible will be coated more heavily than substrates placed off to the side.

Evaporation is considered a line of sight technique due to the high sticking probability of the evaporated species. A number of works in the 1950s and 1960s focused on measuring the sticking probabilities of condensed metals such as gold, silver, or

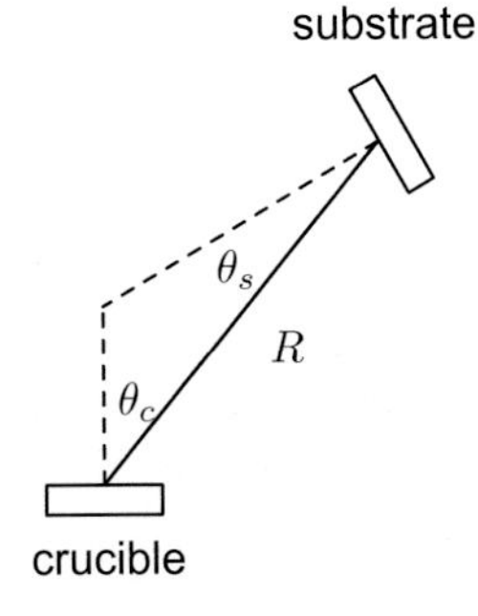

Fig. 2.1 The growth rate in evaporation depends on both the distance and the angle between the crucible and the substrate

copper, on themselves. The results show sticking probabilities ranging from 0.6 to 1.0 [1–6]. Many of these results were based on measured growth rates, and consequently in some of the works the resulting sticking probabilities refer to the net mass transfer process, which also accounts for the possible evaporation of the substrate as well as nucleation effects.

The high sticking probabilities coupled with the high directional flow compromise the step coverage of films deposited using evaporation methods, leading to shadowing effects in steps on the substrate, as shown in the simulated profile in Fig. 1.9. One way of mitigating the effects of a directional flux is to rotate the substrate inside the evaporator. This at the very least makes the flux axially uniform with respect to the normal to the substrate surface.

2.1.2 Sputtering

The widespread use of sputtering as a deposition technique in semiconductor manufacturing has lead to a number of studies to evaluate the conformality of films deposited by this technique, as well as to the development of different approaches to enhance its step coverage. In this section we will focus on the application of sputtering to conformality. While a brief description of this technique is given in Sect. 2.1.2.1, the reader is referred to the many texts and review articles covering the fundamental sputtering and its many applications [7, 8].

2.1.2.1 Fundamentals of Sputtering

In sputter deposition, ion bombardment is used to vaporize atoms from the surface of a target. Momentum transfer from the incident ions causes a collision cascade that leads to the vaporization of atoms near the surface. These atoms are subsequently transported to the surface.

The ions causing the sputtering originate from a high electron density plasma generated near the target. One of the most common configurations is the magnetron, in which strong cross electric and magnetic field help confine the plasma close to the target. Also, due to the physical nature of the process, a wide range of materials can be sputter deposited: in addition to metals, insulating materials can be deposited using RF sputter deposition.

The angular distribution of the ejected atoms can change with the sputtering conditions: while in many cases it is close to a cosine distribution, it also has been shown to change from under-cosine to over-cosine with increasing ion energy. The ejected atoms are not thermalized, and instead they are characterized by an energy distribution function with a long tail that can extend into 10 s of electronvolts.

The need to sustain a discharge implies that the pressure chamber during sputtering is higher than during evaporation. Gas-phase collisions cause scattering of the ejected atoms, affecting the overall transport efficiency of the process, and contribute to the thermalization of the species sputtered from the target. A further variation of sputtering commonly referred to as *reactive sputtering* carries out this process with a background pressure of a reactive gas such as oxygen or nitrogen to tune the composition of the resulting film.

As in the case of evaporation, measured and calculated sticking probabilities yield high values, [9–11] which makes it hard for sputtering to grow conformal films on high surface area materials.

2.1.2.2 Conformality of Sputtering

The combination of the higher pressure and a higher energy of incident species tend to increase the step coverage of sputtered films with respect to the evaporated counterparts. However, the sticking probabilities are still high. This makes it hard for sputtering to achieve the coating of high aspect ratio or high surface area materials.

One of the factors that can contribute to the enhancement of conformality in sputtering is the presence of high energy ions: when these ions get to the surface they cause sputtering of the film itself, followed by redeposition elsewhere on the substrate, as schematically shown in Fig. 2.2. The typical profiles of the growth in vertical features such as trenches and vias are characterized by a reduction in thickness in the sidewalls due to the effect of this ion bombardment, while the thickness of the bottom surface is increased due to a reduction of the ion flux and the effect of redeposition.

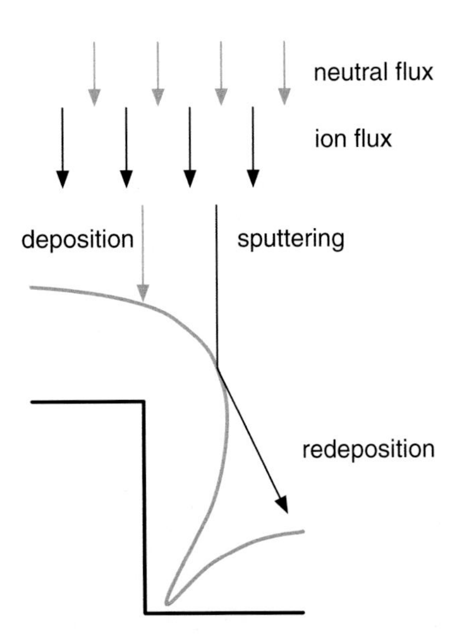

Fig. 2.2 A flux of ions towards the substrate can induce sputtering and redeposition, improving the step coverage achieved in films deposited by sputtering

2.1.3 Approaches to Improve Step Coverage in PVD Methods

Since PVD techniques have poor step coverage, there has been a lot of interest in developing ways of increasing their conformality. Here we provide a quick overview of some of the methods described in the literature.

Promoting surface and bulk diffusion One way to improve the step coverage of metal films in both evaporation and sputtering is to increase the substrate temperature to enhance the mobility of species. The key here is that by enhancing diffusion atoms can be drawn into the via following the gradient in surface potential created by the surface curvature. This is a method that, for instance, has been successfully used in the sputtering of aluminum [12].
A reduced pressure also enhances surface mobility by reducing the pinning of adatoms by gas phase impurities.

Bias sputtering The application of a bias increases the energy of the incident particles up to a point in which part of the material can be sputtered from the surface. Re-sputtering of the deposited film will increase the conformality of the material at the expense of lower growth rates [13].

Force fill techniques Here the approach is to sputter films with extremely poor step coverage, typically producing a cusp at the top corners of a trench or via. Eventually, the tops converge pinching off any further deposition, and also sealing the low pressure gas inside a void inside the feature. A high pressure treatment of these films forces the collapse of the metal bridges into the features [7].

Collimated sputtering The use of a collimator with high aspect ratio holes filters out particles reaching the substrate at a high angle. This improves the step coverage by allowing species to reach deeper into the features, although at the cost of a lower deposition rate due to the fraction of species lost in the collimator pores [14].

Ionized PVD Here the ejected atoms are ionized by a second plasma. This allows the control of the angular distribution function of the ionized fraction of the sputtered ions by applying a dc bias. The consequence of this process is a near vertical distribution of incident species that changes the growth profiles inside high aspect ratio trenches and vias. While the thickness at the bottom of the trench is greatly increased with respect to conventional PVD processes, the thickness of the coating on the walls is typically lower [15–17].

2.2 Chemical Vapor Deposition

In chemical vapor deposition (CVD), growth takes place through the reaction or decomposition of gaseous species, forming a solid film on a heated surface as one of the reaction byproducts.

The diversity of CVD processes and operating conditions is extremely high. Growth temperatures vary from room temperature in plasma-enhanced CVD (PECVD) to more than 1500 °C during the epitaxial growth of SiC. Depending on the operating pressure, one can distinguish between atmospheric pressure CVD (APCVD), low-pressure CVD (LPCVD), and ultrahigh vacuum CVD (UHPCVD). The differences in pressure do not just translate into differences in equipment: as the pressure goes down, the relative importance of homogenous versus heterogeneous processes is progressively reduced. Changes in pressure also affect the mass transport inside the reactor. These changes greatly impact the deposition process.

Except at ultrahigh vacuum conditions, where flows at a reactor scale are in the collisionless or Knudsen regime, in CVD species reaching a surface can be considered to be in local thermodynamic equilibrium and they are typically characterized by a Maxwellian distribution function. Consequently, the incident fluxϕ of a given species to the surface, defined as the number of atoms per unit area and time impinging on the surface, is given by its equilibrium value:

$$\phi = \frac{p}{\sqrt{2\pi M k_B T}} \tag{2.2}$$

where T is the temperature, k_B is Boltzman's constant, M is the molecular mass, and p is the partial pressure of that particular species.

It would be impossible to cover the range of CVD processes available within the scope of this book. Instead, the reader is directed to the reviews published in the literature [18, 19].

2.2.1 Impact of Kinetics on the Conformality of CVD Processes

Process conditions and reactor configuration have a strong influence on the conformality of a CVD process, in many cases in non-trivial ways [20]. Changes in temperature and pressure can have a dramatic effect on the ability of a process to grow conformally on high aspect ratio features. In Table 2.1 we present a selection of CVD processes that have shown a promising ability to coat, and in some cases fully infiltrate, high aspect ratio and high surface area materials.

The fundamental criterion to achieve conformality in CVD is the existence of a *weak dependence of the growth rate with precursor pressure*. That way, the impact of precursor depletion inside a high aspect ratio or nanostructure materials on the growth rate is minimized, thereby improving its step coverage.

While the range of processes available is extremely large, we can use simple models to understand the dependence of growth rate with pressure and therefore the impact of processing conditions on conformality. In the next sections we present two of such models, involving one and two different precursors.

Table 2.1 Summary of conformal CVD processes

Material	Reactant 1	Reactant 2	Ref.	Comment
Cu	bis hexafluoroacetonate copper (II)	H_2	[21]	H_2 as carrier gas, additional solvent plays a role
	Cu(hfac)vtms	-	[22]	H_2 used as carrier gas
HfB_2	$Hf(BH_4)_4$	-	[23]	Conformal coating of 19:1 aspect ratio trench
HfO_2	tetrakis diethylamido hafnium	O_2	[24, 25]	Step coverage actually improves with temperature
Mo	$Mo(CO)_6$	-	[26]	Photonic crystal infiltration
Ru	Ru(C5H5)2	O_2	[27]	Filling of contact hole of aspect ratio 4
RuO2	$Ru(tmhd)_3$	O_2	[28]	Temperature dependence on step coverage of low aspect ratio via
SiO_2	$Si(OC_2H_5)_4$	-	[29]	Infiltration of alumina membrane
TaN	tertbutylimido trisdiethylamido tantalum	-	[30]	Conformal growth in high aspect ratio trenches
Ta_2O_5	$Ta(OC_2H_5)_5$	O_2	[31]	LPCVD, step coverage comparison with reactive sputtering
TiN	$TiCl_4$	NH_3	[32]	Complete filling of low aspect ratio trench
TiN	tetrakis diethylamido titanium	NH_3	[33]	APCVD, step coverage as a function of growth rate
W	$W(CO)_6$	-	[26]	Photonic crystal infiltration
WN_x	bis-tertbutylimido bis-tertbutylamido tungsten	-	[34]	Single source precursor

2.2.1.1 Single Precursor Processes: First Order Langmuir Kinetics

One of the simplest models that is able to reproduce the dependence with temperature and pressure found in many single-source precursor systems is the first order Langmuir kinetics. Given a gas phase species $A_{(g)}$, this model assumes that the overall reaction takes place through a surface intermediate $A_{(s)}$:

$$A_{(g)} + X_{(s)} \rightleftharpoons A_{(s)} \tag{2.3}$$

$$A_{(s)} \rightarrow X_{(s)} + \text{Film} + \text{Byproducts} \tag{2.4}$$

Consequently, the growth process can be conceptually divided into two steps: a reversible adsorption-desorption process followed by an irreversible reaction with the surface. Here X_s represents an available surface site for adsorption. The growth rate GR in atoms per surface area is then given by the rate of Eq. 2.4, which in turn depends on the surface concentration of the intermediate species $A_{(s)}$. If we denote by Θ the fraction of occupied sites, so that $\Theta = [A_{(s)}]s_0$, where s_0 is the average area of a surface site, the growth rate can be expressed as:

$$\mathrm{GR} = k_r \Theta / s_0 \tag{2.5}$$

where k_r is the surface reaction rate coefficient, and the growth rate has units of atoms per unit time and unit surface area.

The fraction of intermediate states on the surface is the result of the balance between adsorption, desorption, and reaction. The adsorption is simply given by the number of particles per unit site, which in this case is given by:

$$\alpha p = s_0 \phi \tag{2.6}$$

where ϕ flux of species per unit area to the surface. The desorption rate is denoted by k_d. In absence of reaction, the balance of these two processes results on a Langmuir adsorption isotherm for the species A. In presence of the irreversible surface reaction pathway we have that:

$$\alpha p(1 - \Theta) - k_d \Theta - k_r \Theta = 0 \tag{2.7}$$

and consequently:

$$\Theta = \frac{\alpha p}{\alpha p + k_r + k_d} \tag{2.8}$$

so that the growth rate is given by:

$$\mathrm{GR} = \frac{k_r}{s_0} \frac{\alpha p}{\alpha p + k_r + k_d} \tag{2.9}$$

We can use Eq. 2.9 together with the definition of sticking probability to calculate β as a function of temperature and pressure:

$$\beta(T, p) = \frac{k_r(T)}{\alpha p + k_r(T) + k_d(T)} \tag{2.10}$$

where the dependence with both p and T has been made explicit.

There are several conclusions that can be extracted from these two equations:

1. We expect a dependence of the conformality of a CVD process with temperature and pressure, since the sticking probability β depends on both temperature and pressure, $\beta = \beta(p, T)$.
2. When the pressure is low enough,

$$\beta \approx \frac{k_r}{k_r + k_d} = \beta_0(T) \tag{2.11}$$

the sticking probability is independent of pressure and it solely depends on temperature as defined by the precursor-surface interaction. The process is said to be in the transport-limited regime, and the growth rate is linear with pressure. Under these conditions, a weak dependence of the growth rate with pressure is achieved when the sticking probability is low.

3. When the pressure is high enough, the pressure term in the denominator in Eq. 2.9 dominates, and the growth becomes pressure independent. This is the regime where we expect to have conformal processes [20, 35].

Within this simple model, the growth rate can be expressed solely as a function of two temperature dependent parameters, $\mathrm{GR}_{\mathrm{sat}}(T)$ and $b(T)$:

$$\mathrm{GR}(T, p) = \mathrm{GR}_{\mathrm{sat}}(T)\frac{b(T)p}{1 + b(T)p} \tag{2.12}$$

The meaning of these two parameters is clear: $\mathrm{GR}_{\mathrm{sat}}$ is the asymptotic value of the growth rate achieved at high pressure and $b(T)$ is a parameter in units of pressure^{-1} that characterizes the pressure at which the growth rate starts to roll over to its saturation value. Figure 2.3 shows how the transition between the low pressure and the high pressure regimes depends on the actual value of the single effective parameter $b(T)$.

This simplified picture works particularly well in cases where growth takes place at low pressures, so that growth is determined by heterogeneous processes. Some examples of single source CVD processes include can be found in Table 2.1. Even if the model provides an oversimplified view of the surface kinetics, many systems share some of the main trends depicted above: these include the transition from a surface reaction limited regime at low temperatures to a gas transport or flux limited regime at high temperatures, followed in many cases by a decrease in the growth rate at even higher temperatures, due to enhanced precursor desorption and/or upstream precursor consumption [36].

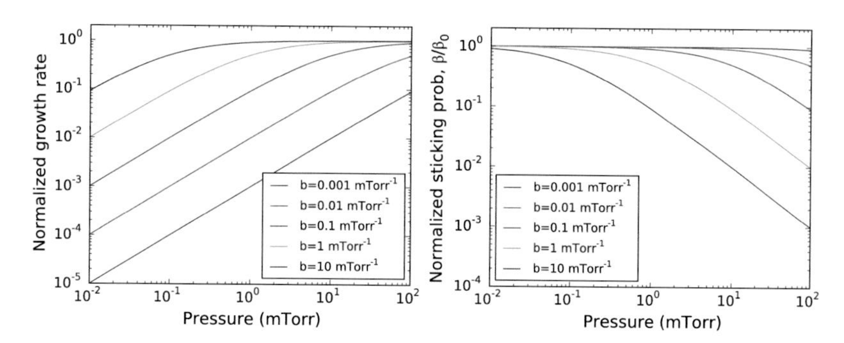

Fig. 2.3 Dependence of growth rate with pressure for a single precursor model

2.2.1.2 Two Reactant Processes

In this case, the CVD process involves more than one precursor, with the second precursor typically helping reach the right composition or phase. This is one of the most common situations in the literature, with many of the examples in Table 2.1 falling under this category.

While in general the kinetics of these processes can be extremely complicated (and also pressure dependent), here we present a simple model that can help illustrate the impact that processing conditions can have in their growth rate and conformality: let us consider a surface composed of two different sites A(s) and B(s). This model assumes that precursors A and B can only react with the surface sites B(s) and A(s), respectively. Therefore, we can view the surface kinetics as a reversible transition between the two different sites at the surface:

$$\mathrm{A(s)} \rightleftharpoons \mathrm{B(s)} \tag{2.13}$$

One example of such model is the surface kinetics proposed for the SiC CVD: different kinetic models for SiC CVD published in the literature can be reduced to this simple scheme [37].

Here we are going to focus on the simplest case in which we have only two gas phase species A and B. The reaction is then given by two irreversible reactions:

$$\mathrm{A} + \mathrm{B(s)} \rightarrow \mathrm{A(s)} + \mathrm{B(b)} \tag{2.14}$$

and

$$\mathrm{B} + \mathrm{A(s)} \rightarrow \mathrm{B(s)} + \mathrm{A(b)} \tag{2.15}$$

where A(s), B(s) and A(b), B(b) are the surface and the bulk species, respectively.

If Θ_A and Θ_B are the fractional coverages of A(s) and B(s) on the surface (with $\Theta_A + \Theta_B = 1$), and k_a and k_b the corresponding reaction rates, then the steady state coverages will be given by:

$$\Theta_A k_b p_B - \Theta_B k_a p_A = 0 \tag{2.16}$$

And the corresponding growth rate in atoms per surface area is given by:

$$\mathrm{GR} = k_a p_A \theta_B = k_b p_B \theta_A = \frac{k_a p_A k_b p_B}{k_a p_A + k_b p_B} \tag{2.17}$$

Just as in the single precursor case, as the pressure of one of the two precursors increases, the growth rate rolls over to reach an asymptotic value. If we assume that precursor B is in excess, then the growth rate will be approximately given by:

$$\mathrm{GR} = k_a p_A \tag{2.18}$$

Under this condition, growth rate is independent of p_B: the precursor A is said to be in a transport or supply limited regime. This condition take place whenever $k_b p_B \gg k_a p_A$.

Many processes involving two precursors are run in excess of one of the two processes. For instance, in the growth of GaN by MOCVD using trimethylgallium and ammonia, the process typically takes place in an excess of ammonia, which leads to a linear growth rate with the trimethylgallium pressure [38].

A more general discussion on the impact of surface kinetics in conformality is provided in Chap. 4.

2.2.2 Strategies to Improve Conformality in CVD

The key to achieve highly conformal, or even superconformal, growth in CVD is to identify a process with the right surface kinetics. Alternatively, it is in principle possible to tweak the surface kinetics of an existing process in such a way that it improves its ability to grow inside high surface area materials. In this section we highlight some of the approaches and systems in the literature that have lead to an improvement of conformality:

Growth inhibition In this approach, the pressure of a secondary species is tuned to inhibit growth, effectively reducing the sticking probability during LPCVD [39]. A simple kinetic model is described in Chap. 4.

Pulsed CVD Some groups have explored the impact of pulsing one of the precursors on film conformality. Examples in the literature include Cu and W deposition [40, 41].

Surfactant-mediated growth This approach is based on the use of surfactants to catalyze the growth. This has been demonstrated for both Cu and Mn CVD using iodine as a catalyst [42, 43]. In this approach, the growing surface is exposed to iodine prior to the growth. The presence of iodine accelerates the growth, and it always remains on the surface due to its high surface mobility. The key to achieve bottom up filling is that as growth proceeds the the area of exposed surface inside the substrate becomes smaller, which leads to an increase in the local concentration of iodine, which in turns accelerates the growth. Using this technique, the authors were able to demonstrate *superconformal deposition* and bottom-up filling of high aspect ratio features.

Competitive adsorption In Sect. 2.2.1.2, we introduced a surface kinetics model in which two precursors reacted with different sites on the surface. Here, two precursors compete for the same surface sites, with one of them effectively blocking the other, resulting on a growth rate that *decreases* with pressure. This mechanism has been proposed to explain MgO superconformal deposition [44, 45]. In Chap. 4 we expand on the possible kinetic models behind this approach to highly conformal deposition.

2.3 Atomic Layer Deposition

Atomic layer deposition is a chemistry based deposition technique that has the unique property of being intrinsically conformal: ideal ALD processes are able to evenly coat high surface area materials, no matter their shape and their total area. There are a number of excellent reviews on this technique and its application to coat or functionalize high surface area materials [46–52]. Like in the previous cases, the goal of this section is to provide the minimum information required to understand the coating and modification of nanostructured or high surface area materials using this technique.

2.3.1 Introduction

In the typical ALD process, a surface is exposed to alternate pulses of two or more different gases. In each pulse, the gas reacts with the surface in a self-limited way: there is a finite number of sites on the surface and, after all the surfaces have been consumed, the reaction stops. In a simple ALD process involving two precursors, such as that shown in Fig. 2.4a, pulses are separated by a purge time, which is intended to avoid the simultaneous presence of A and B in the gas phase. Consequently, an ALD process involving a precursor and a co-reactant can be described as a sequence of four different times t_1–t_2–t_3–t_4, each corresponding to one step of a single ALD cycle: t_1 corresponds to the dose time of precursor 1, t_2 is the purge time after precursor 1, t_3 is the dose time of precursor 2, and t_4 is the purge time of precursor 2. This set of four times constitutes an ALD cycle, which is repeated as many times as needed in order to reach the desired thickness. The self-limited nature of ALD persists only if the purge times are large enough to prevent cross-talk between the two precursors. In Fig. 2.4b, for instance, we have a process where precursors A and B overlap with time: this would affect the self-limited nature of the resulting process, since it would

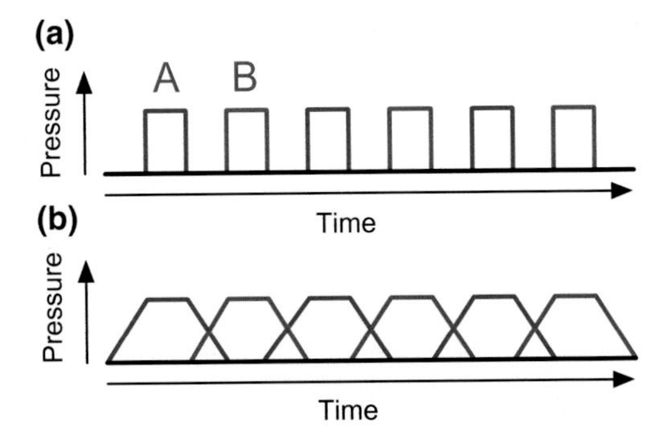

Fig. 2.4 Schematic representation of the pulsed nature of an ALD process. In (**a**) precursors A and B do not overlap in time, as required in a well-behaved process. In (**b**), A and B overlap, potentially leading to spurious CVD

allow A and B to react simultaneously with the surface or with each other through gas phase collisions.

One of the main consequences of the sequential nature of ALD is that, if other process can be characterized in terms of a growth rate, ALD processes are characterized by their *growth per cycle*, defined as the thickness achieved in a single ALD cycle.

The range of materials accessible to ALD include a wide range oxide, nitrides, sulfides, fluorides, chalcogenides, and metals comprising more than 60 different elements. In Fig. 2.5 we summarize the range of ALD processes available for oxides, nitrides, sulfides, and elemental species.

2.3.2 Models of ALD Surface Kinetics

While the mechanistic aspects of ALD are by no means trivial, [46, 53] the simplest model that capture the main features of an ALD dose is an irreversible first order Langmuir surface kinetics: let us consider a surface that has a finite number of surface sites, so that each site is characterized by an average surface area s_0. We can define the *surface coverage fraction* Θ as the fraction of those sites that are occupied by a reacted molecule. The simplest model is then to assume that the reaction probability β has a first-order dependence on the fraction of available sites on the surface $1 - \Theta$,

$$\beta = \beta_0(1 - \Theta) \tag{2.19}$$

so that the change in coverage as a function of time is given by:

$$\frac{d\Theta}{dt} = s_0 \frac{1}{4} \bar{v} n \beta_0 (1 - \Theta) \tag{2.20}$$

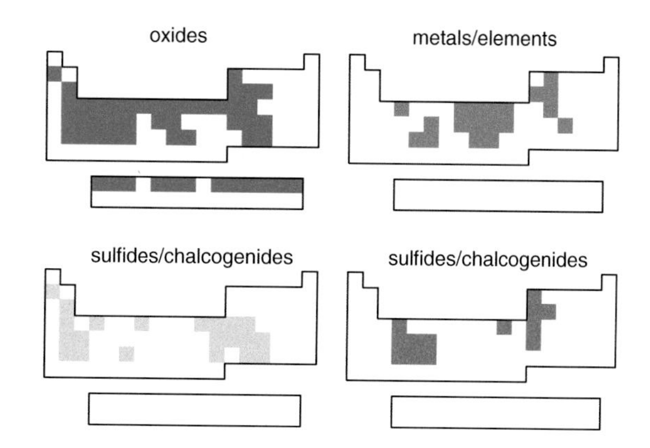

Fig. 2.5 Range of oxide, metal/elements, nitrides, sulfides and chalcogenide materials available by ALD, as reported in Ref. [52]

Here β_0 is the *bare* reaction probability, or the probability that an incident molecule reacts with a pristine surface.

Equation 2.20 can be integrated to obtain the evolution of the surface coverage with time:

$$\Theta(t) = 1 - \exp\left(-s_0 \frac{1}{4}\bar{v}\beta_0 n t\right) \tag{2.21}$$

As the exposure increases, the surface coverage $\Theta(t) \to 1$. The ALD process is said to have reached saturation.

Since $\beta = \beta_0(1 - \Theta)$, this implies that the sticking probability evolves with time as:

$$\beta(t) = \beta_0 \exp\left(-s_0 \frac{1}{4}\bar{v}\beta_0 n t\right) \tag{2.22}$$

And the surface becomes non-reactive ($\beta \to 0$) asymptotically as the exposure increases. This is the definition of a self-limited process. In Fig. 2.6 the evolution of the surface coverage with surface flux per unit site, defined as:

$$N = s_0 \frac{1}{4}\bar{v} n t \tag{2.23}$$

is shown for different values of the bare sticking probability β_0. Large (small) values of β_0 correspond to fast (slow) reaction kinetics.

If we define a certain coverage value c_0, we can use Eq. 2.21 to determine the dose time t needed to reach that coverage for a certain precursor pressure p:

$$t_0 = \frac{4k_B T}{s_0 p \bar{v} \beta_0} \left|\log(1 - c_0)\right| \tag{2.24}$$

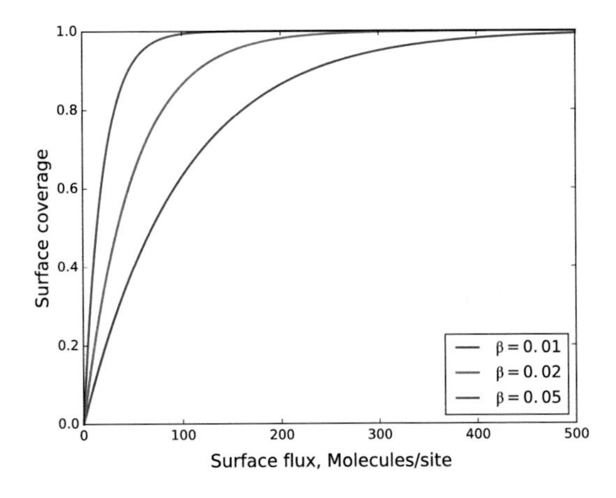

Fig. 2.6 Evolution of the surface coverage with precursor flux for three values of the sticking probability: low sticking probabilities increase the exposures required to reach saturation

If c_0 is our saturation criterion, then t_0 represents the saturation time of our precursor. It is evident that this time is inversely proportional to both the precursor pressure and the bare reaction probability β_0. Dose times in an ALD process have to be larger than this value in order to ensure that saturation is achieved everywhere in the reactor.

We can generalize this model to consider two precursors A and B, and we denote Θ_A and Θ_B as the fraction of the surface covered by A and B species, respectively. Obviously, $\Theta_A + \Theta_B = 1$. In an analogous way to Eq. 2.20, the evolution with time of the surface coverage of species A will be ruled by the simple equation:

$$\frac{d\Theta_A}{dt} = s_0 \frac{1}{4} \bar{v}_A n_A \beta_A \Theta_B - s_0 \frac{1}{4} \bar{v}_B n_B \beta_B \Theta_A \quad (2.25)$$

Here we are neglecting the impact of gas phase collisions: we are assuming that despite A and B being present simultaneously present in the gas phase, the kinetics is driven by heterogeneous reactions.

From Eq. 2.25 we can see how the presence of a background pressure of B affects the evolution of the sticking probability. Simplifying the notation in Eq. 2.25, so that:

$$\frac{d\Theta_A}{dt} = \alpha \Theta_B - \gamma \Theta_A \quad (2.26)$$

and taking into account that the sticking probability of precursor A is given by:

$$\beta(t) = \beta_A \Theta_B \quad (2.27)$$

we obtain that:

$$\beta(t) = \beta_A \frac{\gamma}{\alpha + \gamma} + \beta_A \frac{\alpha}{\alpha + \gamma} e^{-(\alpha + \gamma)t} \quad (2.28)$$

And for large exposures we obtain that:

$$\beta(t) \to \beta_A \frac{\gamma}{\alpha + \gamma} \quad (2.29)$$

As shown in Fig. 2.7, if in the case of a pure ALD process the sticking probability tends asymptotically to zero, when there is an overlap between the two precursors the sticking probability tends asymptotically to a constant value that is different from zero: the simultaneous presence of A and B in the gas phase gives leads to a spurious CVD component. Growth is no longer self-limited.

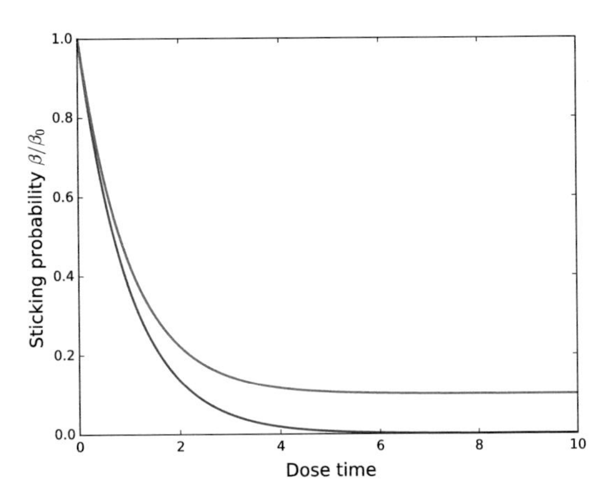

Fig. 2.7 Evolution evolution of the sticking probability with time for a purely self-limited (*blue*) and a process with a non self-limited component (*green*) due to the overlap of the two precursors in the gas phase

2.3.3 Application of ALD to High Surface Area Materials

From Eq. 2.28 it is clear that, even in the case of an ideal self-limited ALD process, the ability to conformally coat a high surface area or nanostructured material will depend on the two precursors not being present simultaneously at any point inside the pores. Consequently, the sequence of dose and purge times t_1–t_2–t_3–t_4 chosen to grow the material has a great impact on the coating of high surface area materials by ALD. Too short times will prevent saturation to be achieved everywhere inside the material. If purge times are too short, cross-talk between the two precursors will affect the overall self-limited nature of the ALD process and lead to thicker films as demonstrated in the previous section.

Two expressions are available in the literature to estimate dose times or exposures in ALD for high surface area materials. The best-known was developed by Gordon *et al* and provides the exposure required to coat circular pores as a function of their aspect ratio (AR): [54]

$$pt = \frac{\sqrt{2\pi M k_B T}}{s_0}\left[1 + \frac{19}{4}\mathrm{AR} + \frac{3}{2}(\mathrm{AR})^2\right] \tag{2.30}$$

This expression was later generalized to account for other types of high surface area materials and situations in which the sticking probability β_0 is low, so that: [55–57]

$$t = t_0 + \frac{L^2}{D\gamma} \tag{2.31}$$

where t_0 is the saturation time on a flat substrate, L is the thickness of the high surface area material, D is the diffusivity of the material, and γ is the so called *excess number*, defined as:

$$\gamma = \frac{p}{k_B T}\frac{s_0}{\bar{s}} \tag{2.32}$$

where $\bar{s}$ is the specific surface area of the material, that is, the surface area per unit volume. In the case of a circular pore of diameter d, the Knudsen diffusivity D is given by:

$$D = \frac{1}{3}\bar{v}d \tag{2.33}$$

A more detailed presentation on the derivation of these expressions is given in Chap. 4.

The consequence of underdosing depends on the microstructure of the high surface area material and the bare sticking probability β_0 of the ALD process. Deendoven *et al* showed that for large values of β_0 saturation in a high surface area material propagates like an step function deeper into the features. However, when β_0 is small, saturation tends to take place more evenly inside the high surface area materials. Under these conditions, doses that fall short do not lead to significant gradients in film thickness, but instead the growth is undersaturated everywhere inside the feature [58]. In Sect. 4.4.1.2 we will show that they parameter controlling the transition between these two behaviors is the Thiele's modulus, which was defined in Chap. 1 as the ratio between the reaction rate and the diffusion rate inside the material.

The propagation of the saturation coverage as a step function in some of these systems can be leveraged to engineer materials so that they are coated with stripes of different compositions. This was demonstrated for the case of anodised aluminum oxide substrates, where the long pores were coated with stripes of different materials by sequentially dosing two or more ALD precursors, with each of the pulses providing doses that fell well short from those required to fully saturate the pore. Successive exposures would react only with the remaining available sites, which were located deeper into the pores, until no more sites were available. This methodology can be employed for instance to selectively grow a stripe of material at a controlled depth inside the pore [59].

Finally, the non-ideality of many ALD processes also affects the ability to coat high surface area materials. Some of these non-idealities include the present of a soft-saturating behavior, the presence of an superimposed CVD component, the potential inhibition of reaction byproducts, and the existing of competing recombination processes, such as in the case of plasma-assisted, and some ozone and peroxide-based ALD processes.

2.4 Summary

In this chapter we have presented an introduction to the different vapor phase deposition techniques and their application to the coating of nanostructured and high surface area materials. In each of them we have emphasized those aspects that are more relevant to the conformality of the deposition process, including the different approaches that can be used to improve their ability to infiltrate nanostructured and high surface area substrates.

References

1. L. Bachmann, J.J. Shin, J. Appl. Phys. **37**(1), 242 (1966)
2. S. Chandra, G.D. Scott, Can. J. Phys. **36**(9), 1148 (1958)
3. J.P. Hirth, G.M. Pound, Prog. Mater Sci. **11**, 1 (1963)
4. R.A. Rapp, J.P. Hirth, G.M. Pound, Can. J. Phys. **38**(5), 709 (1960)
5. R.A. Rapp, G.M. Pound, J.P. Hirth, J. Chem. Phys. **34**(1), 184 (1961)
6. H. Schwarz, J. Appl. Phys. **37**(12), 4341 (1966)
7. S.A. Campbell, *Fabrication Engineering at the Micro and Nanoscale* (Oxford University Press, 2008)
8. S.M. Rossnagel, J. Vac. Sci. Technol. A **21**, S74 (2003)
9. B. Emmoth, H. Bergsaker, Nucl. Instrum. Methods Phys. Res., Sect. B **33**(1–4), 435 (1988)
10. M.R. Weller, K.M. Hubbard, R.A. Weller, D.L. Weathers, T.A. Tombrello, Nucl. Instrum. Methods Phys. Res., Sect. B **42**(1), 19 (1989)
11. J.D. Kress, D.E. Hanson, A.F. Voter, C.L. Liu, X.Y. Liu, D.G. Coronell, J. Vac. Sci. Technol. A **17**, 2819 (1999)
12. M. Inoue, K. Hashizume, H. Tsuchikawa, J. Vac. Sci. Technol. A **6**(3), 1636 (1988)
13. Y. Homma, S. Tsunekawa, J. Electrochem. Soc. **132**(6), 1466 (1985)
14. S.M. Rossnagel, D. Mikalsen, H. Kinoshita, J.J. Cuomo, J. Vac. Sci. Technol. A **9**(2), 261 (1991)
15. S. Hamaguchi, S.M. Rossnagel, J. Vac. Sci. Technol. A **14**, 2603 (1996)
16. S.M. Rossnagel, J. Vac. Sci. Technol. A **16**, 2585 (1998)
17. S.M. Rossnagel, J. Hopwood, J. Vac. Sci. Technol. B **12**, 449 (1994)
18. M.J. Hampdensmith, T.T. Kodas, Chem. Vap. Deposition **1**(1), 8 (1995)
19. K.L. Choy, Prog. Mater Sci. **48**(2), 57 (2003)
20. J.J. Hsieh, J. Vac. Sci. Technol. A **11**, 78 (1993)
21. B. Zheng, E.T. Eisenbraun, J. Liu, A.E. Kaloyeros, Appl. Phys. Lett. **61**, 2175 (1992)
22. A. Jain, T.T. Kodas, R. Jairath, M.J. Hampdensmith, J. Vac. Sci. Technol. B **11**, 2107 (1993)
23. S. Jayaraman, Y. Yang, D.Y. Kim, G.S. Girolami, J.R. Abelson, J. Vac. Sci. Technol. A **23**, 1619 (2005)
24. Y. Ohshita, A. Ogura, A. Hoshino, S. Hiiro, T. Suzuki, H. Machida, Thin Solid Films **406**, 215 (2002)
25. Y. Ohshita, A. Ogura, A. Hoshino, T. Suzuki, S. Hiiro, H. Machida, J. Cryst. Growth **235**, 365 (2002)
26. P. Nagpal, D.P. Josephson, N.R. Denny, J. DeWilde, D.J. Norris, A. Stein, J. Mater. Chem. **21**(29), 10836 (2011)
27. T. Aoyama, M. Kiyotoshi, S. Yamazaki, K. Eguchi, Japn. J. Appl. Phys. Part 1 **38**, 2194 (1999)
28. J.M. Lee, J.C. Shin, C.S. Hwang, H.J. Kim, C.G. Suk, J. Vac. Sci. Technol. A **16**(5), 2768 (1998)
29. S.C. Yan, H. Maeda, K. Kusakabe, S. Morooka, Y. Akiyama, Ind. Eng. Chem. Res. **33**, 2096 (1994)
30. M.H. Tsai, S.C. Sun, C.P. Lee, H.T. Chiu, C.E. Tsai, S.H. Chuang, S.C. Wu, Thin Solid Films **270**, 531 (1995)
31. H. Shinriki, M. Nakata, IEEE Trans. Electron Devices **38**, 455 (1991)
32. N. Yokoyama, K. Hinode, Y. Homma, J. Electrochem. Soc. **138**(1), 190 (1991)
33. J.N. Musher, R.G. Gordon, J. Electrochem. Soc. **143**(2), 736 (1996)
34. M.H. Tsai, S.C. Sun, H.T. Chiu, S.H. Chuang, Apl. Phys. Lett. **68**(10), 1412 (1996)
35. A. Yanguas-Gil, Y. Yang, N. Kumar, J.R. Abelson, J. Vac. Sci. Technol. A **27**, 1235 (2009)
36. C.J. Taylor, D.C. Gilmer, D.G. Colombo, G.D. Wilk, S.A. Campbell, J. Roberts, W.L. Gladfelter, J. Am. Chem. Soc. **121**(22), 5220 (1999)
37. A. Yanguas-Gil, K. Shenai, in *ECS Transactions*, vol. 64 (Electrochemical Soc Inc, Pennington, 2014), pp. 133–143
38. R.P. Parikh, R.A. Adomaitis, J. Cryst. Growth **286**(2), 259 (2006)

39. N. Kumar, A. Yanguas-Gil, S.R. Daly, G.S. Girolami, J.R. Abelson, J. Am. Chem. Soc. **130**(52), 17660 (2008)
40. H. Kikuchi, Y. Yamada, A.M. Ali, J. Liang, T. Fukushima, T. Tanaka, M. Koyanagi, Jpn. J. Appl. Phys. **47**, 2801 (2008)
41. K. Kim, K. Yong, Electrochem. Solid-State Lett. **6**, C106 (2003)
42. K.C. Shim, H.B. Lee, O.K. Kwon, H.S. Park, W. Koh, S.W. Kang, J. Electrochem. Soc. **149**(2), G109 (2002)
43. D. Josell, S. Kim, D. Wheeler, T.P. Moffat, S.G. Pyo, J. Electrochem. Soc. **150**, C368 (2003)
44. W.B. Wang, J.R. Abelson, J. Appl. Phys. **116**(19), 194508 (2014)
45. W.J.B. Wang, N.N. Chang, T.A. Codding, G.S. Girolami, J.R. Abelson, J. Vac. Sci. Technol. A **32**(5), 051512 (2014)
46. R.L. Puurunen, J. Appl. Phys. **97**(12), 121301 (2005)
47. M. Knez, K. Niesch, L. Niinisto, Adv. Mater. **19**(21), 3425 (2007)
48. H. Kim, H.B.R. Lee, W.J. Maeng, Thin Solid Films **517**(8), 2563 (2009)
49. G.N. Parsons, S.M. George, M. Knez, MRS Bull. **36**(11), 865 (2011)
50. H.B. Profijt, S.E. Potts, M.C.M. van de Sanden, W.M.M. Kessels, J. Vac. Sci. Technol. A **29**(5), 050801 (2011)
51. C. Detavernier, J. Dendooven, S.P. Sree, K.F. Ludwig, J.A. Martens, Chem. Soc. Rev. **40**(11), 5242 (2011)
52. V. Miikkulainen, M. Leskela, M. Ritala, R.L. Puurunen, J. Appl. Phys. **113**(2), 021301 (2013)
53. F. Zaera, J. Phys. Chem. Lett. **3**(10), 1301 (2012)
54. R.G. Gordon, D. Hausmann, E. Kim, J. Shepard, Chem. Vap. Deposition **9**, 73 (2003)
55. A. Yanguas-Gil, J.W. Elam, ECS Trans. **41**(2), 169 (2011)
56. A. Yanguas-Gil, J.W. Elam, Chem. Vap. Deposition **18**(1–3), 46 (2012)
57. A. Yanguas-Gil, J.W. Elam, J. Vac. Sci. Technol. A **32**(3), 031504 (2014)
58. J. Dendooven, D. Deduytsche, J. Musschoot, R.L. Vanmeirhaeghe, C. Detavernier, J. Electrochem. Soc. **156**, P63 (2009)
59. J.W. Elam, J.A. Libera, M.J. Pellin, P.C. Stair, Appl. Phys. Lett. **91**(24), 243105 (2007)

Chapter 3
Fundamentals of Gas Phase Transport in Nanostructured Materials

The coating of a high surface area or a nanostructured material is driven by the reactive transport of one or more species inside its pores or features. The fundamental processes that molecules undergo are summarized in Fig. 3.1: they can react with the surface or be reemitted, they can scatter or undergo reactions in the gas phase, and if they are mobile they can adsorb and diffuse on the surface.

Models have to capture the impact of these processes in order to accurately predict the conformality of the resulting film. They have to provide information not only of the fraction of incident species that will react with the surface, a term that echoes the concept of utilization factor used in heterogeneous catalysis and introduced in Sect. 1.4, but also on *where* inside the material the species react.

Let us consider the reactive transport of a single species inside a nanostructured or high surface area material. If $\mathbf{x}$ is a point of the surface, then the growth rate at $\mathbf{x}$, $\mathrm{GR}(\mathbf{x})$ will be given by:

$$\mathrm{GR}(\mathbf{x}) = \frac{M_m}{\rho}\beta(\mathbf{x})\phi(\mathbf{x}) \tag{3.1}$$

where $\beta(\mathbf{x})$ is the sticking probability of the incoming species at $\mathbf{x}$, $\phi(\mathbf{x})$ is the incident flux of molecules at $\mathbf{x}$, M_m is the mass that gets incorporated into the film per molecule, and ρ is the mass density of the resulting film.

From Eq. 3.1 it is clear that, in order to understand and predict the growth inside the high surface area material, we need to determined the local incident flux $\phi(\mathbf{x})$ at every point of the surface. Moreover, given $\phi(\mathbf{x})$, and consequently $\mathrm{GR}(\mathbf{x})$, we can also determine how the surface of the nanostructured material will evolve with time, since the surface evolution will be driven locally by the equation:

$$\mathbf{v}(\mathbf{x}) = \mathrm{GR}(\mathbf{x})\hat{\mathbf{n}} \tag{3.2}$$

where $\hat{\mathbf{n}}$ is the normal vector to the surface S at $\mathbf{x}$, as shown in Fig. 3.2. Introducing the different approaches that can be used to determine $\phi(\mathbf{x})$ is the goal of this chapter.

A. Yanguas-Gil, *Growth and Transport in Nanostructured Materials*,
SpringerBriefs in Materials, DOI 10.1007/978-3-319-24672-7_3

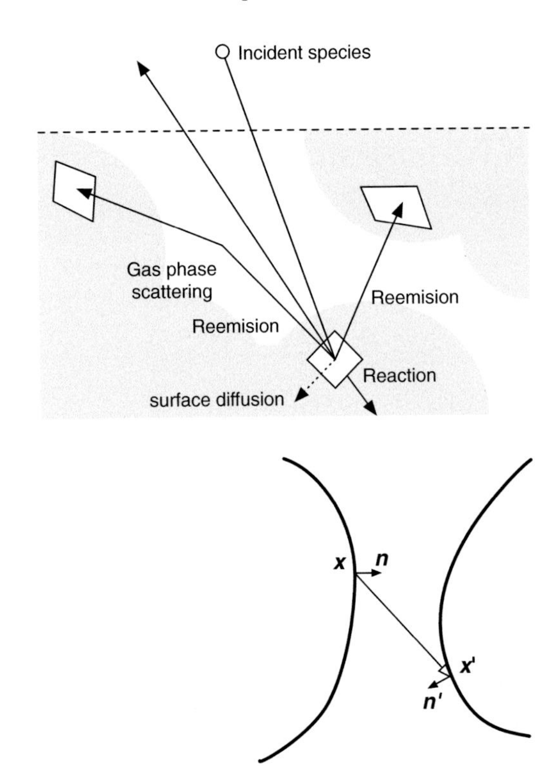

Fig. 3.1 Scheme of the different processes taking place from an atomistic perspective during the coating of high surface area material. All these steps will affect the conformality and the microstructure of the material

Fig. 3.2 In the ballistic transport model, the transition probability between two points $\mathbf{x}$ and $\mathbf{x}'$ of a surface S depends solely on the geometry of the surface and the particle-reemission model

There are essentially two different approaches to model reactive transport within high surface area or porous materials: ballistic models, which accurately model the interaction of molecules with the inner surface of the nanostructured material, and continuum descriptions of the reactive transport that rely on the definition of a diffusion coefficient. This diffusivity is strongly dependent on the structure and porosity of the material.

Both approaches have a long history: ballistic models can be traced back to the seminal work of Clausing involving collisionless flow in circular pipes [1], while the introduction of the concept of diffusivity in the Knudsen flow originates in the work of Knudsen in the early 20th century [2, 3]. A key difference between the two approaches is that ballistic models are typically used to model transport within pores or nanostructures with well-defined geometries, while the continuum description has been used to model transport within both geometrical features and random porous materials. In this chapter we introduce these two approaches.

3.1 Ballistic Models

The problem of growth inside a porous material is characterized by three characteristic time scales: the characteristic time for shape evolution t_f, the characteristic time for surface chemistry change t_s, and the characteristic time for particle transport t_p.

The ballistic model of reactive transport is based on the following assumptions:

1. The transport takes place under Knudsen flow conditions, $\mathrm{Kn} > 1$. The mean free path λ of a molecule is much larger than the characteristic distance d within a feature or high surface area material. Consequently, a particle leaving the surface at $\mathbf{x}$, as shown in Fig. 3.2 will move following a linear trajectory until it arrives at another point of the surface $\mathbf{x}'$. The probability of this $\mathbf{x} \rightarrow \mathbf{x}'$ transition will be solely determined by the re-emission process and the geometry of the material. An important characteristic of Knudsen flow is that the reactive transport process becomes scale independent: the same models that are applied to simulate molecular flow at a reactor scale in ultrahigh vacuum conditions can be applied to model the transport inside nanostructured materials.
2. The transport process takes place under steady-state or quasi steady state conditions. This assumes that the characteristic time for shape evolution, and for changes in the surface chemistry, is much longer than the equilibration time for the local fluxes at the surface, $t_f, t_s \gg t_p$.

3.1.1 Fundamental Equations

Under these assumptions, let $\mathbf{x}$ and $\mathbf{x}'$ be two points inside the high surface area materials, as shown in Fig. 3.2. Let $\phi(\mathbf{x})$ be the flux of species hitting surface at the point $\mathbf{x}$, and $\phi_r(\mathbf{x})$ the outgoing flux. The incident and outgoing fluxes are related through the *sticking probability* β:

$$\phi_r(\mathbf{x}) = (1 - \beta(\mathbf{x}))\phi(\mathbf{x}) \tag{3.3}$$

We can also define the *view factor* $q(\mathbf{x}', \mathbf{x})$ between two points $\mathbf{x}$ and $\mathbf{x}'$ of the feature as the probability that molecules leaving $\mathbf{x}'$ reach $\mathbf{x}$. Similarly, we can define the exit probability $q_e(\mathbf{x}')$ as the probability that a molecule exits the feature. Finally, let us define $q_0(\mathbf{x})$ as the entrance probability, or the probability that a molecule entering the feature has its first collision with a point $\mathbf{x}$ of the material. Since a molecule leaving a particle is going *somewhere*, $q(\mathbf{x}', \mathbf{x})$ and $q_e(\mathbf{x}')$ are related by the constraint:

$$\int d\mathbf{x} q(\mathbf{x}', \mathbf{x}) + q_e(\mathbf{x}') = 1 \tag{3.4}$$

where the integration extends over the whole inner surface of the material. Likewise, since a particle going into the feature has to go *somewhere*, we have necessarily that:

$$\int d\mathbf{x} q_0(\mathbf{x}) \leq 1 \tag{3.5}$$

where the inequality accounts for the possibility that a particle entering the nanostructured material leaves it without colliding with its wall, for instance in the case of the particle crossing a straight pore without undergoing any wall collisions.

With these definitions, the incident flux at $\mathbf{x}$ will be given by the following integral equation:

$$\phi(\mathbf{x}) = q_0(\mathbf{x})\phi_0 + \int d\mathbf{x}' q(\mathbf{x}', \mathbf{x})\phi_r(\mathbf{x}') \tag{3.6}$$

Using Eq. 3.3,

$$\phi(\mathbf{x}) = q_0(\mathbf{x})\phi_0 + \int d\mathbf{x}' q(\mathbf{x}', \mathbf{x})(1 - \beta(\mathbf{x}'))\phi(\mathbf{x}') \tag{3.7}$$

Equation 3.7 constitutes the core of the ballistic model. In the non-reactive case, $\beta(\mathbf{x}) = 0$, and Eq. 3.7 simply transforms into:

$$\phi(\mathbf{x}) = q_0(\mathbf{x})\phi_0 + \int q(\mathbf{x}', \mathbf{x})\phi(\mathbf{x}')d\mathbf{x}' \tag{3.8}$$

Among other applications, Eq. 3.7 has been used to model transport at a reactor scale for ballistic or Knudsen flows. An identical model is also used to simulate radiative transport. This means it is possible to take advantage of methodologies developed to model radiative heat transfer and apply them to our problem. This is for instance the case of the view factors, which have been tabulated for many geometrical configuration in radiative transfer monographs.

In the case of a single pore, when the position z along the pore is the relevant parameter, this equation reduces to

$$\phi(z) = q_0(z)\phi_0 + \int_0^L q(z, z')\phi(z')dz' \tag{3.9}$$

Equation 3.9 is the equation derived by Clausing on his seminal work on the transport of rarified gases on circular pores of a finite length.

From Eq. 3.7 it is clear that the models depends on the following key parameters:

- The particle reemission model, codified in the view factors $q(\mathbf{x}, \mathbf{x}')$ and $q_e(\mathbf{x})$.
- The source gas distribution, that is, the angular distribution of particles reaching to our material from the gas phase, codified in $q_0(\mathbf{x})$.
- The surface reactivity, contained in the sticking probability $\beta(\mathbf{x})$. It is important to note that this coefficient includes not only processes that lead to thin film growth but also others that may lead to a deactivation or transformation of the incoming species into a different species.

In addition to these parameters, there are two other processes that may become relevant under some conditions: surface mobility can become important whenever the mean surface diffusion length becomes of the order of the characteristic size of the feature d. Gas phase collisions and reactions can also take place when the density of reactive species is high. This will impact the value of the view factors $q(\mathbf{x}, \mathbf{x}')$. However, it is important to note that since Eq. 3.7 extends only to the surface of the feature, ballistic models are not ideally suited to track chemical reactions in the gas phase.

3.1.2 Source Gas Distribution

The determination of the entrance probability $q_0(\mathbf{x})$ requires understanding the angular distribution at which molecules reach the surface of the high surface area material. This is particularly important for processes exhibiting a high sticking probability, since then the influence of $q_0(\mathbf{x})$ is much higher. In the limiting case of a line of sight process with a perfectly reacting surface where $\beta(\mathbf{x}) = 1$, the growth rate at every point is directly proportional to $q_0(\mathbf{x})$:

$$\mathrm{GR}(\mathbf{x}) = \frac{M_m}{\rho}\beta(\mathbf{x})q_0(\mathbf{x})\phi_0(\mathbf{x}) \tag{3.10}$$

With localized sources and low enough pressures, the source gas distribution depends on the geometry of the source and the angular distribution function of the emitted species. Here we are going to focus on the equilibrium case, where the source gas distribution is characterized by a cosine distribution.

3.1.2.1 Equilibrium Case: Cosine Distribution

If the gas is close to a thermodynamic equilibrium with the surface then particles will reach the surface following a cosine angular distribution.

This can be shown as follows: if we have an isotropic flux of molecules, the flux per unit solid angle is constant regardless of the orientation of that solid angle:

$$\frac{d\phi}{d\Omega} = \varphi \equiv \text{const} \tag{3.11}$$

If we consider a geometry such as that shown in Fig. 3.3, where $\hat{\mathbf{n}}$ is the normal vector to the surface and $\hat{\mathbf{r}}$ represents a unit vector pointing in the direction of the solid angle element $d\Omega$, the flux of molecules crossing an area element dA will be given by:

$$dN = \frac{d\phi}{d\Omega}\hat{\mathbf{n}} \cdot \hat{\mathbf{r}} d\Omega dA \tag{3.12}$$

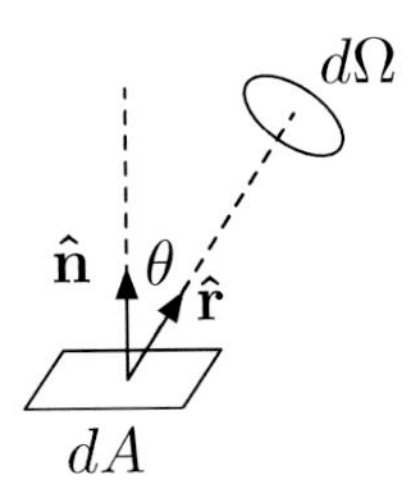

Fig. 3.3 The flux per unit surface area coming from a solid angle element $d\Omega$ is proportional to the cosine of the angle connecting the surface normal $\hat{\mathbf{n}}$ and the unit vector $\hat{\mathbf{r}}$ connecting dA and $d\Omega$. This is the origin of the cosine distribution when the flux is isotropic

or

$$dN = \varphi \cos\theta d\Omega dA \tag{3.13}$$

so that if $d\phi$ represents the flux per unit surface area and unit solid angle:

$$d\phi = \frac{dN}{dA} = \varphi \cos\theta d\Omega \tag{3.14}$$

3.1.3 *Particle Reemission Model*

If the sticking probability is less than one, then particles can undergo multiple collisions inside the high surface area material. The calculation of view factors $q(\mathbf{x}, \mathbf{x}')$ depends on the reemission model that is used to model the particle-surface interaction.

There are two limiting cases: in a first scenario the particle has an elastic interaction with the surface and undergoes a specular reflection. Alternatively, the particle can be thermalized during the interaction with the surface, losing memory of its incident velocity. In this case, the particle undergoes diffuse scattering. These two cases were contemplated already in 19th century by Maxwell. The bulk of the models in the literature assume a diffuse scattering model, since elastic scattering is more prevalent when the energy transfer process taking place between the incident molecule and the surface is not efficient, for instance when the incident molecules are not thermalized or when they have only a few internal degrees of freedom We will therefore focus on the case of diffuse scattering.

In the thermodynamic limit, the emission of a particle from a surface follows a cosine distribution. The main argument for this is that under equilibrium conditions the velocity distribution function of the molecules in a gas just above the surface should be the same as in the case where the surface is not present. Therefore, the same isotropic flux that we expect from a gas in the thermodynamic equilibrium should be expected from the molecules coming from a surface.

If we use the same argument that we used for the incident flux in the previous section, even if the flux per solid angle is isotropic, the flux per unit area will be proportional to the cosine of the angle between the normal of the surface and the direction from which the flux is being measured.

The differential outgoing flux per solid angle will then be given by:

$$d\phi_r = \phi_r \cos\theta \frac{d\Omega}{\pi} \tag{3.15}$$

so that when integrated over $d\Omega$,

$$\int d\phi_r d\Omega = \phi_r \tag{3.16}$$

The cosine reemission law has been experimentally verified in many occasions, including the work by Clausing in 1930 [4]. In 1957, Hurlbut carried out molecular scattering studies on polished steel, polished aluminum, and unpolished glass at different incident angles, finding that the cosine reemission was the prevalent behavior [5].

We can draw a parallelism between surface reemission and radiative emission: surfaces in which reemission is dominated by a diffuse scattering driven by a cosine distribution are the equivalent of a Lambert reflector, which emits radiation with the same intensity regardless of the angle and behaves as a perfectly diffuse surface.

3.1.4 View Factors

So far no assumption has been made regarding the value of $q(\mathbf{x}', \mathbf{x})$ and $q_e(\mathbf{x})$, other that they are constrained by the equalities defined in Eqs. 3.4 and 3.5. In fact, Eq. 3.7 would be equally valid for any value of the Knudsen number as long as the steady state approximation holds. However, Eq. 3.7 is typically applied for $\mathrm{Kn} \gg 1$. In this section, we will therefore focus on view factors determined on the Knudsen flow regime and a cosine re-emission model.

3.1.4.1 General Equation

Let us consider the geometry shown in Fig. 3.4. The differential view factor dF_{12} is defined as the fractional flux of particles leaving dA_1 that reaches dA_2. Based on the cosine reemission law introduced in the previous section, we will have that the differential view factor will be given by the product:

$$dF_{12} = \cos\theta_1 dA_1 \frac{d\Omega_2}{\pi} \tag{3.17}$$

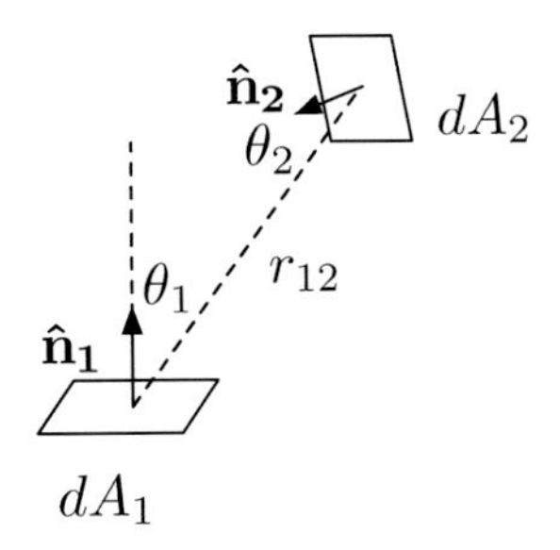

Fig. 3.4 The differential view factor between two surface elements dA_1 and dA_2 are a function of their distance r_{12} and the angles between their normals and the straight line joining the two elements

where π is introduced as a normalization constant.

The solid angle $d\Omega_2$ is simply given by:

$$d\Omega_2 = \frac{dA_2 \cos\theta_2}{r_{12}^2} \tag{3.18}$$

where r_{12} is the distance between the two area elements and θ_2 is the angle between the line joining the two surface elements and the normal to dA_2.

This leads to the following expression for the differential view factor:

$$dF_{12} = \frac{\cos\theta_1 \cos\theta_2}{\pi r_{12}^2} dA_1 dA_2 \tag{3.19}$$

Equation 3.19 assumes that there are no obstacles in the path of the molecule between the two surface elements causing shadowing. In a general case, this is not true, and we have to modify this expression by adding a *visibility factor* κ_{12} between the two elements, so that:

$$dF_{12} = \kappa_{12} \frac{\cos\theta_1 \cos\theta_2}{\pi r_{12}^2} dA_1 dA_2 \tag{3.20}$$

Now the view factor is zero if another part of the feature is obstructing the path between the two elements.

From Eq. 3.20, the view factor $q(\mathbf{x}, \mathbf{x}')$ can be expressed as:

$$q(\mathbf{x}, \mathbf{x}') = \kappa(\mathbf{x}, \mathbf{x}') \frac{\cos\theta \cos\theta'}{\pi r^2} \tag{3.21}$$

where $r = |\mathbf{x} - \mathbf{x}'|$ represents the distance between $\mathbf{x}$ and $\mathbf{x}'$.

One important consequence of the relation (3.21) is that it makes Eq. 3.7 invariant to isotropic dilations, that is, the solution of Eq. 3.7 becomes independent of the absolute size on the feature. If $d\mathbf{x}$ and $d\mathbf{x}'$ are two area elements centered in $\mathbf{x}$ and $\mathbf{x}'$, respectively, the probability that a particle leaving $d\mathbf{x}$ reaches $d\mathbf{x}'$ is given by: $q(\mathbf{x}, \mathbf{x}')d\mathbf{x}'$. If we now assume an arbitrary change of scale $x \rightarrow ax$, then it is easy to see that $d\mathbf{x}' \to a^2 d\mathbf{x}'$ and $r^2 \to a^2 r^2$, resulting in both terms cancelling each other in Eq. 3.7.

This scale invariance holds as long as the Knudsen number is higher than one. When this is not the case, the reemision probability change, and the coating becomes dependent on the actual size of the feature. This property is the reason why in trenches and features one traditionally uses the concept of aspect ratio AR.

3.1.4.2 View Factor Calculations

There is a wealth of view factors for different geometrical configurations that have already been calculated in the context of radiative transfer models. Here we focus on two cases: infinite rectangular trenches and circular pores and vias.

Infinite trenches: For the case of infinite trenches a useful view factor is the one connecting two parallel infinite slabs as shown in Fig. 3.5. If we consider the first slab centered at the origin and aligned with the y axis and normal vector $\hat{\mathbf{n}} = \hat{\mathbf{e}_x}$, and a second slab also aligned with the y axis and located at a distance $r = \sqrt{x^2 + z^2}$ from the origin and with normal vector $\hat{\mathbf{n}}'$ given by:

$$\hat{\mathbf{n}}' = \cos\theta \hat{\mathbf{e}_x} + \sin\theta \hat{\mathbf{e}_z} \tag{3.22}$$

the view factor between the two slabs will be:

$$q(x,z) = \frac{-x^2\cos\theta - xz\sin\theta}{2\left(x^2+z^2\right)^{3/2}} \tag{3.23}$$

We can particularize Eq. 3.23 for the two relevant situations in a rectangular trench showed in Fig. 3.6. The wall-to-wall view factor between two points of a trench of width d separated a distance z is given by:

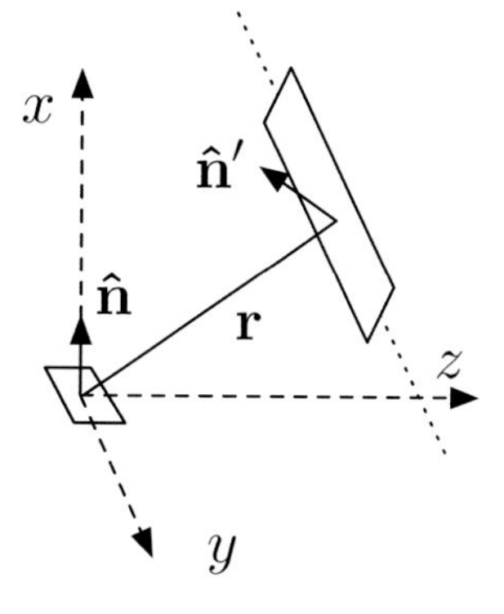

Fig. 3.5 Relevant parameters required for the calculation of the view factor for two parallel infinite slabs

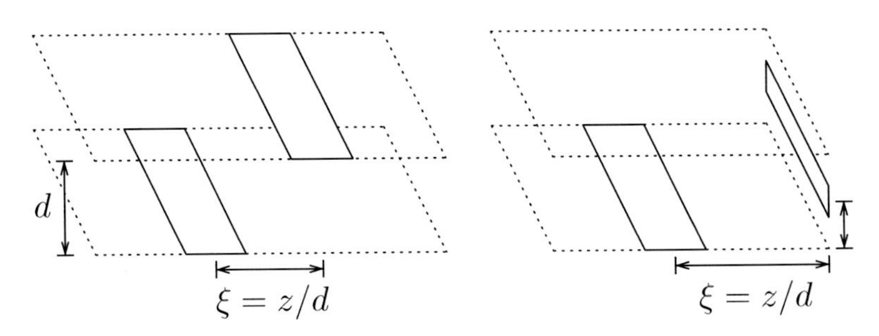

Fig. 3.6 Scheme of the two geometrical configurations that are needed to calculate the view factors in a rectangular trench

$$q(z) = \frac{d^2}{2\left(d^2 + z^2\right)^{3/2}} \tag{3.24}$$

where we just considered $\theta = -\pi$, and $x = d$ in Eq. 3.23. If we then define $\xi = z/d$ and $q(\xi)$ as $q(\xi)d\xi = q(z)dz$, we can normalize this expression with respect to the trench width:

$$q(\xi) = \frac{1}{2}\frac{1}{\left(1 + \xi^2\right)^{3/2}} \tag{3.25}$$

where now ξ extends from 0 to AR $= L/d$. Note that the derivation of this expression takes into account the contribution of one of the two sidewalls.

Likewise, the differential view factor of a strip at the entrance of bottom of the trench from a point in the wall can be obtained by considering $\theta = -\pi/2$. The differential view factor will be given by:

$$q(z, x) = \frac{xz}{2\left(x^2 + z^2\right)^{3/2}} \tag{3.26}$$

where z is the distance from the wall to the top/bottom of the feature, and x is the height of the strip with respect to the wall plane.

Conversely, normalizing with respect to the trench width d and defining $\eta = x/d$ and $q(\xi, \eta)$ so that $q(z, x)dx = q(\xi, \eta)d\eta$, we obtain that:

$$q(\xi, \eta) = \frac{\eta\xi}{2\left(\eta^2 + \xi^2\right)^{3/2}} \tag{3.27}$$

We can now use this equation to obtain the transmission probability from the mouth to the wall by integrating over the whole mouth of the trench and considering the contribution of the two walls:

$$q_0(\xi) = 2\int_0^1 d\eta \frac{\eta\xi}{2\left(\eta^2 + \xi^2\right)^{3/2}} = 1 - \frac{\xi}{\left(1 + \xi^2\right)^{1/2}} \tag{3.28}$$

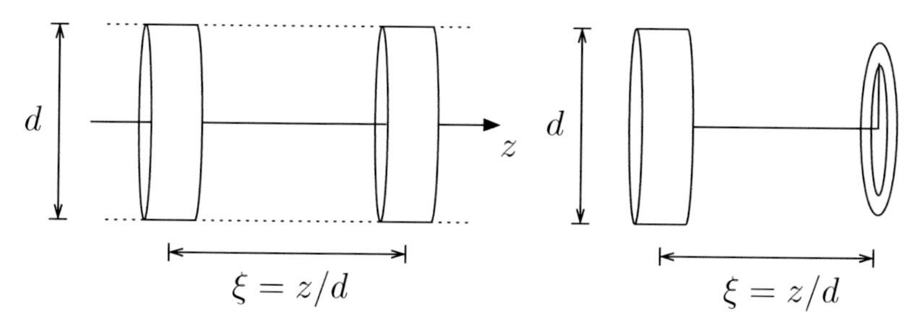

Fig. 3.7 Wall-wall view factors for a circular via: resulting expressions depend only on the normalized depth $\xi = z/d$, where d is the diameter of the pore

It is easy to see that Eq. 3.28 is normalized correctly, since:

$$\int_0^\infty q_0(\xi)d\xi = 1 \tag{3.29}$$

That is, the probability that a particle entering an infinitely deep trench collides with the wall is one. Note that in order to properly normalize the results on a trench, view factors must take into account contributions from the two walls.

Circular pores: Just as in the case of rectangular trenches, it is possible to determine analytic expression for the view factors. In this case, the two configurations are the wall to wall and wall to base configurations shown in Fig. 3.7.

If r is the pore radius, the differential view factor between two rings located a distance z apart will be given by:

$$q(z) = \frac{1}{d}\left(1 - \frac{3}{2}\frac{|z|}{\left(z^2+d^2\right)^{1/2}} + \frac{1}{2}\frac{|z|^3}{\left(z^2+d^2\right)^{3/2}}\right) \tag{3.30}$$

If $\xi = z/d$, where d is the pore diameter, we can obtain a normalized equivalent to Eq. 3.30:

$$q(\xi) = 1 - \frac{3}{2}\frac{|\xi|}{\left(1+\xi^2\right)^{1/2}} + \frac{1}{2}\frac{|\xi|^3}{\left(1+\xi^2\right)^{3/2}} \tag{3.31}$$

It is easy to see that Eq. 3.31 is properly normalized, since:

$$\int_{-\infty}^\infty q(\xi)d\xi = 1 \tag{3.32}$$

Likewise, for the wall-base view factor, we consider a circular ring of radius r' located at a distance z from the inner surface. Then, the view factor will be given by:

$$q(z, r') = 2zr\frac{r^2+z^2-r'^2}{\left[(r^2+r'^2+z^2)^2 - 4r^2r'^2\right]^{3/2}} \tag{3.33}$$

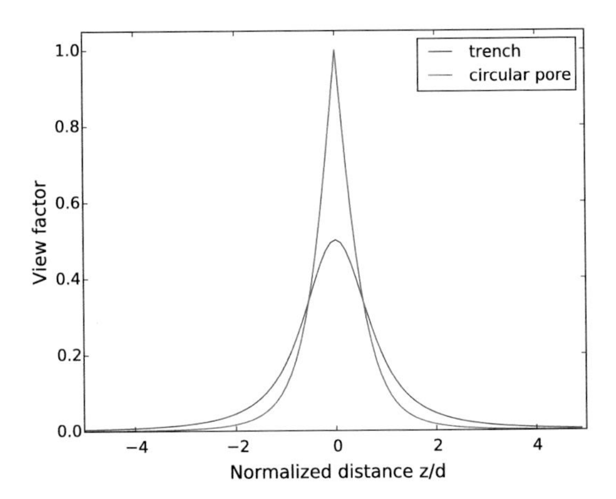

Fig. 3.8 Wall-to-wall view factors for a rectangular trench and a circular via as a function of z/d, where d is the diameter of the via or the width of the trench

When we integrate Eq. 3.33 between 0 and the pore radius we obtain the view factor between the inner walls and an opening at a distance z:

$$q_o(z) = \frac{1}{2r}\left[\frac{z^2 + 2r^2}{\sqrt{z^2 + 4r^2}} - z\right] \tag{3.34}$$

It is easy to see that $q_o(0) = 1/2$, that is, the probability that a particle just by the opening of the pore abandons the pore is 50 %. By normalizing Eq. 3.34 so that $\xi = z/d$, with d the pore diameter, we obtain that:

$$q_o(\xi) = \frac{2\xi^2 + 1}{2\sqrt{1 + \xi^2}} - \xi \tag{3.35}$$

The view factors described above constitute the basis of ballistic simulations in the literature in trenches and vias.

Figure 3.8 shows a comparison between the wall-to-wall view factor for a rectangular trench and a circular via, illustrating the impact that the cross-section of a pore has on the ability to transport molecules. It is clear from Fig. 3.8 that the transport along a trench is easier than along a circular via for the same characteristic size.

3.2 Single Particle Approaches to Ballistic Transport

If molecule-molecule interactions can be neglected inside a nanostructured material, the transport inside features of pores can be treated as a single particle process. In this section we introduce two different approaches that take advantage of the single-particle condition to solve the reactive transport inside a feature. These are the use of kinetic Monte Carlo simulations and Markov chain models.

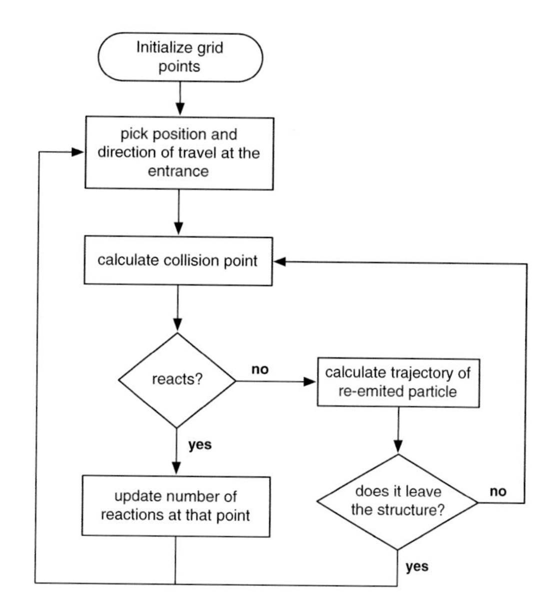

Fig. 3.9 Example of a kinetic Monte Carlo algorithm to model growth inside a feature

3.2.1 *Kinetic Monte Carlo Simulations*

Kinetic Monte Carlo simulations represent an alternative approach to particle transport. One of the earliest applications was the work of Davies in 1960 to model molecular flow on ducts and elbows of different shapes [6].

A typical algorithm used to determine growth profiles using a kinetic Monte Carlo is shown in Fig. 3.9. This is based on the model of Wulu et al. [7] This algorithm tracks the collision of particles inside a feature until they either react with the surface or leave the porous material. This process is repeated a desired number of times until enough statistical data are accumulated.

This method offers two key advantages:

1. It can consider other situations that are not easy to model using deterministic approaches. For instance, when gas phase collisions are considered, the deterministic calculation of the reemission probabilities becomes extremely complicated, while they can be more easily incorporated through Monte Carlo approaches.
2. It sidesteps the problem of having to calculate the view factors, instead calculating the collision of each particle with the feature *on the fly*. This can be a significant advantage if the features have complex shapes. A brute force algorithm to calculate n view factors scales with n as $O(n^3)$, where the cube term comes from the need to determine shadowing effects. While this can be reduced under highly symmetric conditions, if the number of elements is high enough it can quickly become computationally very expensive, particularly if the shape of the feature evolves with time and the view factors need to be recalculated periodically.

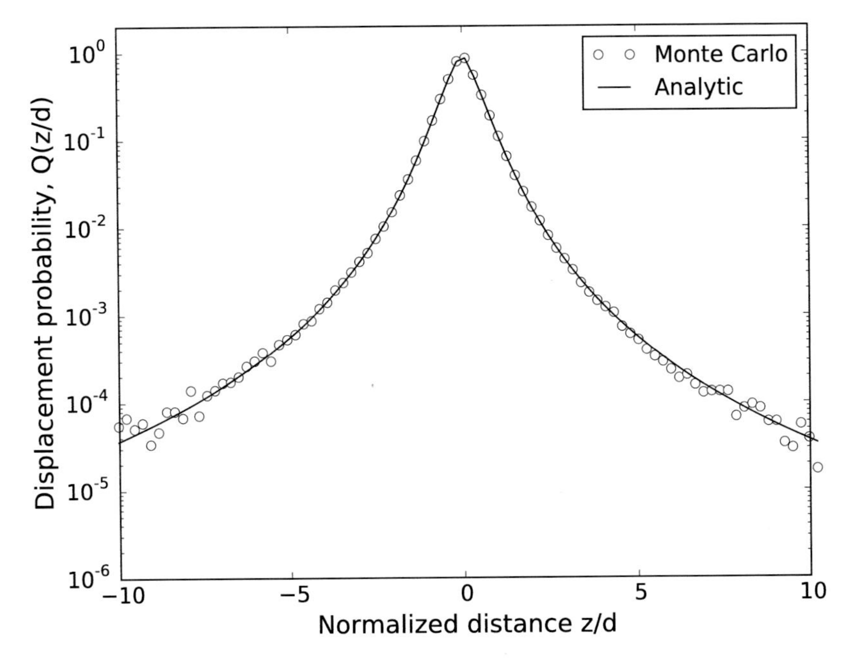

Fig. 3.10 Comparison between the analytical view factor and the view factors calculated using a kinetic Monte Carlo simulation for a cylindrical pore. Simulation results are obtained from 1,000,000 independent trajectories

By accumulating a sufficiently high number of trajectories one can statistically sample the material and extract view factors directly using kinetic Monte Carlo simulations. Figure 3.10 shows a comparison between the view factors derived as a result of 1,000,000 trajectories using a Monte Carlo approach and the analytical expression given by Eq. 3.31 for a cylindrical pore. Depending on the geometry, the trajectory of a single particle can be calculated really fast: for instance, to treat ballistic transport inside an elbow, Davis decomposes the structure into two cylinders, and reduces the problem to determining which of the two cylinders the molecule collides with, something that is achieved by a mere comparison between the distances for the intersection with each cylinder and the dividing plane [6].

The main advantage of kinetic Monte Carlo simulations is that the determination of the next collision of a single particle process is usually faster than if all the view factors need to be calculated. Monte Carlo codes can also be more easily adapted to different geometries. However, kinetic Monte Carlo simulations also have some disadvantages: firstly, a large number of individual trajectories are required to get good statistics. Secondly, it also becomes very inefficient if the sticking probability is low. Under these conditions, the particles will undergo many collisions before they eventually react with the surface.

3.2.2 *Markov Chain Formulation*

An alternative approach is to cast the reactive transport inside a porous material as a Markov chain. Diffuse scattering makes particle reemission a memoryless process: once a particle has arrived to a point $\mathbf{x}$ inside the feature, the reemission probability is codified in a view factor $q(\mathbf{x}, \mathbf{x}')$ that is independent of the history of that particle prior to reaching $\mathbf{x}$. This memoryless feature is the fundamental requirement of a Markov chain.

Let $p_0(\mathbf{x})$ be the probability that a particle is initially in a point $\mathbf{x}$ of the surface. The probability that a particle is in $\mathbf{x}'$ after the first collision $p_1(\mathbf{x}')$ can be obtained from $p_0(\mathbf{x})$ through the view factor:

$$p_1(\mathbf{x}') = \sum_{\mathbf{x}} p_0(\mathbf{x})q(\mathbf{x}, \mathbf{x}') \tag{3.36}$$

where the sum is extended to all the points of the surface.

More generally, the probability after $n + 1$ collisions can be expressed as:

$$p_{n+1}(\mathbf{x}') = \sum_{\mathbf{x}} p_n(\mathbf{x})q(\mathbf{x}, \mathbf{x}') \tag{3.37}$$

This equation proves that the ballistic transport of species inside a nanostructured or high surface area materials satisfy the Markov property. One of the advantages of casting the transport process as a Markov Chain is that we can take advantage of a well-developed formalism to determine the probabilistic outcome of our reactive transport process: where a particle is going to react when it penetrates inside a high surface area material and with which probability. This section will follow closely the approach introduced in Ref. [8]. However, some of the notation has been changed to make it consistent with the rest of the chapter.

3.2.2.1 Reactive Transport as a Markov Chain

The main difference between the reactive transport and the simple non-reactive example mentioned above is that to model the reactive transport of species inside a high surface area material we need to consider two different kind of events. The first group is formed by events involving particle reemission, which essentially transition the system from one state to the next. The second group includes all the different end game scenarios of the reactive transport process: the irreversible reaction with the surface and the molecule abandoning the feature without reacting.

We will therefore consider a Markov chain with two different types of states: *transient states* and *absorbing states*. Absorbing states are states that represent end-game scenarios of the Markov chain. Once the system reaches an absorbing state it stays there forever. For every point of the feature we therefore have a state i representing the particle reaching the surface, but we also need to consider a second

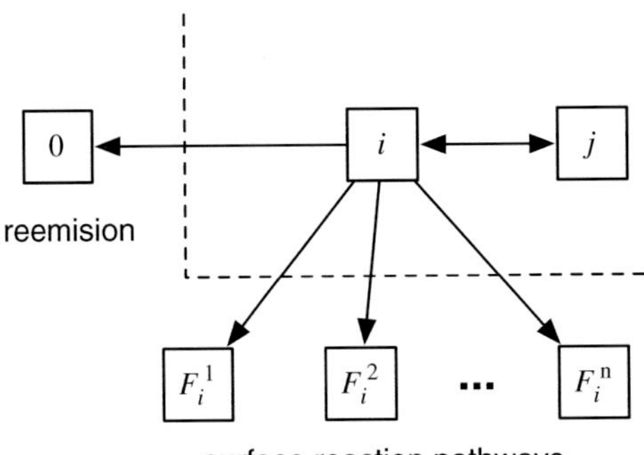

Fig. 3.11 When reactive transport is casted as a Markov chain, a molecule reaching a section of the inner surface of a high surface area material, represented by a transient state i, can undergo a number of possible transitions: it can be reemitted to another point of the feature, represented by a transient state j, it can irreversibly react with the surface through one of potentially many reaction pathways to reach an absorbing state F_i^k, or it can leave the feature, reaching the absorbing state 0. Reproduced from Ref. [8] with permission

state F_i representing the irreversible reaction of the surface. Therefore, we can model reactive transport as three different type of transitions:

- Reemission from a point i to a point j of the surface, represented as the $i \rightarrow j$ transition.
- Reaction of the molecule at a point i of the surface, represented as the $i \rightarrow F_i$ transition.
- A particle leaving the feature from point i, represented as a $i \rightarrow 0$ transition.

This is summarized in Fig. 3.11, where the possibility of reaching many different absorbing states at the surface, each representing a separate irreversible reaction pathway, is considered.

Each of the potential transitions enumerated above is characterized by a transition probability. If $P(i \rightarrow j)$, $P(i \rightarrow F_i)$, and $P(i \rightarrow 0)$ represent the corresponding transition probabilities to the three potential scenarios covered in this section, it is clear that:

$$\sum_j P(i \rightarrow j) + P(i \rightarrow F_i) + P(i \rightarrow 0) = 1 \tag{3.38}$$

We can now use the definitions introduced in Sect. 3.1 to populate these probabilities. The probability that the particle reacts is simply the sticking probability at i,

$$P(i \rightarrow F_i) = \beta_i \tag{3.39}$$

The probability that a particle transitions to another point j of the feature is determined by the probability that it *does not* react with the surface and the view factor defined in Eq. 3.21:

$$P(i \rightarrow j) = (1 - \beta_i)q_{ij} \tag{3.40}$$

Likewise, the probability that the particle escapes from the feature without reacting is given by:

$$P(i \rightarrow 0) = (1 - \beta_i)q_{i0} \quad (3.41)$$

The initial probability p_0 is simply given by the view factors from the entrance of the feature:

$$p_0(i) = q_{0i} \quad (3.42)$$

with $p_0(0) = p_0(F_i) = 0$.

Therefore, if $p_n(\sigma)$ represents the probability that a molecule finds itself in the state σ, where σ belongs to $\{\ldots, i, F_i \ldots, 0\}$, after n collisions, we can model the reactive transport as a sequence of p_n where p_n and p_{n+1} are related by the transition probability matrix P whose coefficients are given by the expressions provided above.

3.2.2.2 Exact Solution of an Absorbing Markov Chain

In the previous section we recast a lot of what was developed in Sect. 3.1 for ballistic transport in the context of Markov chains. In this section we show how to extract the probabilistic outcome of the reactive transport based on the theory of absorbing Markov chains.

One of the properties of absorbing Markov chains is that they can be solved exactly: the probability matrix **P** can always be expressed in the so-called standard form. The standard form is obtained when states are sorted in such a way that all absorbing states are listed before the transient states of the system. This allows us to represent the transition probability matrix **P** in the following block form:

$$\mathbf{P} = \begin{bmatrix} \mathbf{I} & \mathbf{0} \\ \mathbf{R} & \mathbf{Q} \end{bmatrix} \quad (3.43)$$

Here **I** is the identity matrix, representing the transition probabilities between absorbing states; **Q** is the matrix containing the transition probabilities between transient states; and **R** is the matrix containing the transition probabilities linking transient to absorbing states.

Based on these matrices, the outcome of the system can be obtained as a function of the initial probabilities π_γ. The probability that the system ends in an absorbing state α is given by:

$$P(\alpha) = \pi_\alpha + \sum_\delta \pi_\delta (\mathbf{MR})_{\delta\alpha} \quad (3.44)$$

where **M** is the matrix defined as:

$$\mathbf{M} = (\mathbf{I} - \mathbf{Q})^{-1} \quad (3.45)$$

where the sum in δ is extended to all the transient states.

We can now take Eq. 3.44 and obtain the probability $P(i)$ that a particle either reacts at a point i of the high surface area material:

$$P(i) = \sum_j \pi_j (\mathbf{MR})_{jF_i} = \beta_i \sum_j q_{0j} \mathbf{M}_{ji} \tag{3.46}$$

The effective reaction probability, that is, the probability that an incoming molecule reacts anywhere in the feature, is given by:

$$\bar{\beta} = 1 - \sum_j q_{0j} (\mathbf{MR})_{j0} \tag{3.47}$$

Moreover, by treating the reactive transport as a single particle process, we can extract information on the average number of collisions of a particle inside the nanostructure before it either reacts or leaves the feature. The average number of steps that it takes to reach one of such end-game scenarios is given by:

$$N_c = \sum_{\gamma\delta} \pi_\gamma \mathbf{M}_{\gamma\delta} \tag{3.48}$$

Equations 3.46–3.48 provide all the information needed to characterize the reactive transport process and calculate the growth rates at each point of the high surface area material.

3.3 Continuum Description: Diffusion-Based Models

Continuum descriptions of the transport of gases through porous media have been extensively used in areas ranging from chemical engineering to soil science, and several reviews are available in the literature [9]. In the context of reactive transport and growth inside nanostructured materials in thin film applications, continuum models have been primarily used to describe the transport of species in chemical vapor infiltration, [10] with some examples also used in the literature for both CVD and ALD [11–14].

These approaches rely on the concept of diffusivity D to model the transport inside a porous material using a diffusion equation. If the diffusivity D is independent on the density of species n, then the transport is given by a simple diffusion equation:

$$\frac{\partial n}{\partial t} - D \frac{\partial^2 n}{\partial z^2} = 0 \tag{3.49}$$

If $D = D(n)$, then the more general form:

$$\frac{\partial n}{\partial t} - \frac{\partial}{\partial z}\left(D\frac{\partial n}{\partial z}\right) = 0 \tag{3.50}$$

must be used.

It is important to note that there are at least two different ways of defining the diffusivity of a species: The self-diffusivity or *tracer diffusivity* is defined in terms of the mean square displacement of molecules, which increases linearly with time, with the proportionality constant related to the self-diffusivity as:

$$\langle r^2(t) \rangle = 6Dt \tag{3.51}$$

This is the well-known Einstein equation.

In contrast, the *transport diffusivity* is defined through the relationship between the molecular flow and the concentration gradient, so that:

$$J = -D(n)\frac{\partial n}{\partial z} \tag{3.52}$$

Both descriptions have been shown to be completely equivalent under Knudsen flow conditions, [15] but that it is not necessarily the case in presence of intermolecular interactions.

One of the advantages of the continuous description is that it encompasses the whole range of Knudsen numbers, from Knudsen flow conditions where the mean free path is larger than that of the mean pore size $\bar{d}$, to viscous flow. This is due to the fact that the assumption of uncorrelated displacements at a molecular level is valid for both collisions with other molecules and with pore walls.

In this section we will focus on the problem of how to determine the diffusivity D under different flow conditions: Sect. 3.3.1 introduces the concept of diffusivity under Knudsen flow conditions and its application to the transport inside porous materials; Sect. 3.3.2 deals with transitional flows when the pore diameter becomes of the order of the mean free path of the molecules; Sects. 3.3.3 and 3.3.4 focus on transport inside micropores and polymers, both of which require independent consideration. Finally, we briefly consider the role that surface adsorption has on the transport of species inside high surface area materials in Sect. 3.3.5.

3.3.1 Knudsen Diffusion Coefficient

3.3.1.1 Knudsen Diffusivity in Circular Pores

The use of a continuum description to gas transport through a porous material goes back to the seminal work of Knudsen [3]. When the mean free path λ is larger than a pore diameter, Knudsen showed that the diffusive flux within a circular capillary of diameter d could be expressed as

$$J_c = -D_0 \frac{dn}{dz} \tag{3.53}$$

where

$$D_0 = \frac{1}{3}\bar{v}d \tag{3.54}$$

Here $\bar{v}$ is the mean molecular velocity given by the kinetic theory of gases as:

$$\bar{v} = \left(\frac{8k_B T}{\pi M}\right)^{1/2} \tag{3.55}$$

If we solve the steady state problem given by Eq. 3.49 using the diffusivity defined in Eq. 3.54, we obtain that for a gas density n_0 on the left side and vacuum on the right side of the pores the equilibrium solution is given by

$$n(z) = n_0 \left(1 - \frac{z}{L}\right) \tag{3.56}$$

and the flow per unit area within the pore is given by:

$$N = -D\frac{\partial n}{\partial z} = -\frac{1}{3}\bar{v}n_0\frac{d}{L} \tag{3.57}$$

If we normalize N with respect to the flux per unit surface area ϕ:

$$\phi = \frac{1}{4}\bar{v}n_0 \tag{3.58}$$

we can define the transmission probability W as the ratio between the flow through the pore and the total incident pore, resulting:

$$W = \frac{N}{\phi} = \frac{4d}{3L} \tag{3.59}$$

This value corresponds with the asymptotic value for long circular tubes. The error between Eq. 3.59 and more exact computations has been shown to be of the order of 1 % for an aspect ratio (L/d) of 50 [16].

3.3.1.2 Extension to Porous and High Surface Area Materials

Equation 3.54 can be generalized to model the transport within a porous material under Knudsen flow conditions, so that the flow can be expressed in terms of the Knudsen diffusion coefficient as follows:

$$J_K = -D_K \nabla n \tag{3.60}$$

Typically D_K is expressed in terms of D_0 as:

$$D_K = D_0 \frac{\varepsilon}{\tau} \tag{3.61}$$

where ε represents the porosity of the material, and τ is a tortuosity factor for the Knudsen flow that represents the fact that in a porous material the trajectories that independent molecules have to follow are larger than in a straight perpendicularly oriented pore.

Using Eq. 3.54, this expression can be expanded to:

$$D_K = \frac{1}{4} \bar{v} \bar{d} \frac{\varepsilon}{\tau} \tag{3.62}$$

In a porous material, the pore diameter d in Eq. 3.62 can be taken as the average pore size. Alternatively, an estimate of the average pore size can be obtained from the pore volume-surface area ratio, so that:

$$d = \frac{4V_t}{S} \tag{3.63}$$

where we are assuming that the pores are equal cylinders of diameter d.

3.3.1.3 Relationship Between Microstructure, Porosity and Tortuosity

The key challenge in modeling transport in porous materials using a diffusive model is to find a expression for the diffusivity of the molecules that reflects the microstructure of the material.

There are a number of experimental data available as well as some simpler models that have been extensively used to model transport through porous materials. One of the simplest models is the random pore model of Wakao and Smith [17]. While originally developed for pelleted materials composed of pressed particles that are themselves porous, here we are going to consider the simple case of non-porous particles.

Their calculation is based on the following argument: let us consider a porous materials composed of pores occupying a volume fraction ε. If we cut a section of this material and we rejoin it at random, the probability of having a continuous path between the two sections will be ε^2. This leads to a simple expression of the diffusivity in terms of the Knudsen diffusion on a cylindrical pore as:

$$D_K = \varepsilon^2 D_0 \tag{3.64}$$

Here the term ε^2 represents the probability to have a continuous path between the two rejoined areas. However, if we compare this expression with Eq. 3.61, we see that this simple model leads to a tortuosity coefficient that is inversely proportional to the porosity of the material:

$$\tau = \frac{1}{\varepsilon} \tag{3.65}$$

In their original work, Wakao and Smith considered a more complex situation in which the particles presented additional microporosity, and extended the model to consider diffusion in the transitional regime where both Knudsen and a molecular diffusion components were present. The latter case will be treated in Sect. 3.3.2. In their work they also compared their model results with experimental data taken on pellets of boehmite particles, showing an excellent agreement [17].

Other authors have proposed different models to determine the tortuosity τ as a function of porosity ε. Some of these have been reviewed by Shen and Chen [18]. Note that in that work the tortuosity is defined in a slightly different way as τ^2 in Eq. 3.61. All these models show in common a shared trend of a tortuosity that increases with decreasing porosity.

An alternative approach to the determination of the impact that the structure of the porous material has on diffusion under Knudsen flow conditions is the use of Monte Carlo simulations. For instance, Evans and co-workers sidestepped the concept of tortuosity and instead focused on the definition and determination of diffusion coefficients based on the calculation of the transmission probability of porous plugs with different simulated microstructures [19].

If we consider a porous slab of thickness L separating a gas at a pressure p from vacuum (Fig. 3.12), a gas molecule impacting the left side of the plug will either pass through the plug, or return back. The diffusive flux will be the product of the impingement rate, given by ϕ_0 and the fraction of molecules transmitted W:

$$\phi_t = W\phi_0 \tag{3.66}$$

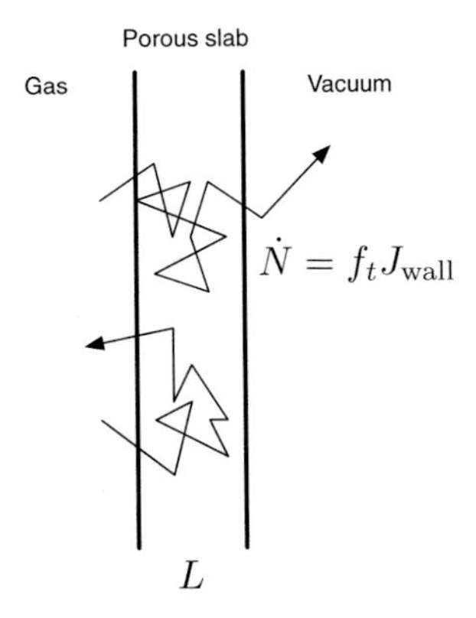

Fig. 3.12 Knudsen flow in terms of the transport through a porous slab

If we now express this flow in terms of a Knudsen diffusivity:

$$\phi_t = D_K \frac{n}{L} \tag{3.67}$$

The Knudsen diffusivity results as follows:

$$D_K = \frac{1}{4}\bar{v}WL \tag{3.68}$$

The first consequence that you can extract from Eq. 3.68 is that for a solid of uniform structure and constant temperature, the transmission probability W should be inversely proportional to L. This is in agreement with the results obtained by Knudsen and Clausing for long circular pores shown in Eq. 3.59.

Second, it provides a simple method to determine the relationship between the tortuosity τ, the porosity and the transmission probability. Comparing Eq. 3.68 with Eq. 3.61 we obtain:

$$\frac{\varepsilon}{\tau} d = \frac{3}{4} WL \tag{3.69}$$

3.3.2 *Transitional Flow*

The transitional flow corresponds to situations in which pore sizes are of the order of the mean free path. Any expression valid in the transition region must asymptotically tend to the diffusivity expressions defined for very high and very low Knudsen numbers. One of the most common approaches is the so-called Bosanquet approximation, that defines a diffusion coefficient given by:

$$\frac{1}{D} = \frac{1}{D_m} + \frac{1}{D_k} \tag{3.70}$$

The interpretation is that wall and molecular collisions are independent events that take place in series. This expression was derived for circular capillaries by Pollard and Present [20].

However, Eq. 3.70 is strictly valid only when the mole fraction y of the diffusing component is very small, that is, in the diluted regime where $y \ll 1$. A more general expression considering two species A and B leads to the following effective diffusion coefficient:

$$\frac{1}{D} = \frac{1 - \alpha y_A}{D_m} + \frac{1}{D_{K_A}} \tag{3.71}$$

where y_A is the molar fraction of species A and α is related to the flux ratio of the two species:

$$\alpha = 1 + \frac{N_B}{N_A} \tag{3.72}$$

In absence of counter flow, or when the molar fraction y_A is very small, Eq. 3.70 is reobtained.

3.3.3 *Diffusion in Micropores*

Knudsen diffusion takes place whenever pore diameter $\bar{d}$ is smaller than the mean free path. However, when pore sizes become of the order of the molecular size, other diffusion modes become important. In this section we review some of them.

3.3.3.1 Surface Diffusion

When pore diameters become small, a second important contribution to mass transport inside nanostructured materials is the surface diffusion component. This diffusion coefficient is thermally activated so that:

$$D_s = D_{s0} \exp\left(-\frac{E_s}{kT}\right) \tag{3.73}$$

The activation energy E_s correlates with the adsorption energy of the gas phase molecules.

Let n be the volumetric density of molecules in the gas phase, and n_s the volumetric density of adsorbed molecules. The total flow will be given by the contribution of gas phase and surface gradients:

$$\dot{N} = -D_g \frac{\partial n}{\partial z} - D_s \frac{\partial n_s}{\partial z} = -\left(D_g + D_s \frac{\partial n_s}{\partial n}\right) \frac{\partial n}{\partial z} \tag{3.74}$$

Assuming that the fraction of adsorbed molecules is in local equilibrium with the gas phase density of molecules inside the pores, the gas phase and fractional surface coverage are related through the adsorption equilibrium constant K_s, so that:

$$K_s = \frac{\partial n_s}{\partial n} \tag{3.75}$$

And the transport process can be modeled considering an effective diffusivity which is the addition of the gas phase and surface contributions:

$$D = \frac{1}{\frac{1}{D_m} + \frac{1}{D_k}} + K_s D_s \tag{3.76}$$

3.3.3.2 Activated Knudsen Diffusion

When the pore diameter becomes of the order of the molecular diameter, the diffusive flow is affected by the interaction potential between the molecules and the walls, and the transport starts becoming affected by the geometrical constraints in the material. An analysis based on the transition state theory predicts that the diffusion will be a thermally activated process due to the need of reaching specific configurations that allow the molecule squeeze through the narrow pores.

3.3.3.3 Single File Diffusion

In some cases, for instance when the pore sizes become of the order of the molecular size, molecules cannot pass each other when moving along a pore. This creates a situation where the movement of molecules is not statistical independent, one of the key requirements to have a normal diffusive behavior. Molecular transport under such conditions have been referred to as single file diffusion [21].

A key characteristic of single file diffusion is that molecular displacements in one direction are more likely to be followed by displacements in the opposite direction. This is due to the fact that after a molecule moves, it leaves behind a gap while its progression can be stopped by other molecules further down the channel, as shown in Fig. 3.13.

Under these conditions, the mean square displacement $\langle z^2 \rangle$ does no longer increase linearly with time, as in the case of conventional or Fickian diffusion, but instead it evolves following the rule [22, 23]:

$$\langle z^2 \rangle = 2Ft^{1/2} \tag{3.77}$$

where:

$$F = \lambda^2 \frac{1-\Theta}{\Theta} \frac{1}{\sqrt{2\pi\tau}} \tag{3.78}$$

where Θ is the site occupancy, λ is the mean displacement per jump and τ is the characteristic time for jumps among diffusion.

Note that Eq. 3.78 does not reduce to the Knudsen value when the occupancy fraction $\Theta \ll 1$. The reason for this is that Eq. 3.78 is derived for conditions where there is a substantial fraction of molecules within the pores. In fact, it has been demonstrated that Eq. 3.78 can be easily derived if one considers the transport of

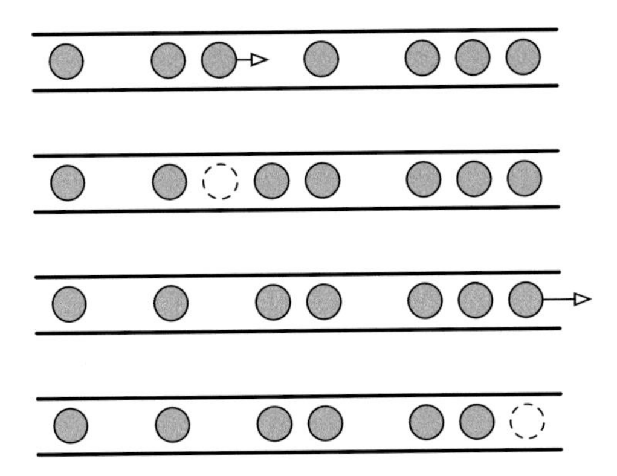

Fig. 3.13 Single file diffusion takes place whenever molecules cannot pass each other inside long pores. It is characterized by two main features: displacements in one direction are more likely to be followed by displacements in the opposite directions, and the overall transport rate may be limited by the desorption rate at the end of the pores

vacancies instead of the molecules, since then it is easy to establish the statistical independence of the displacement of the different vacancies within the pore [23]. Consequently, other methods, such as kinetic Monte Carlo simulations, are needed to capture the transition between the Knudsen and single file diffusion regimes.

A second key mechanism taking place under single file diffusion conditions is the exchange of molecules with the surroundings at the end of the pores. Each adsorption/desorption process causes a shift in the mean distribution of species in a pore and, when biased towards one of the extremes, provides an additional component to the diffusion constant that can be the dominant contribution for long times [24].

3.3.4 Diffusion in Polymers

Polymers are the limiting case of a high surface area material, with several groups in the literature showing the ability of some precursors such as alkyl and halide precursors to react with functional groups within the polymers in a self-limited way.

The characteristic time for infiltration within polymer films is much larger than the characteristic exposure times in atomic layer deposition. This is due to the fact that rather than a porous medium, the permeability of a polymer films toward molecules in the gas phase depends on both the solubility and the mobility of the species inside the polymer.

The permeability of polymers to different gases, defined as the product of the solubility and diffusion coefficient, is typically limited by the low values of molecular diffusion within polymers: the rate controlling process in the diffusion of gas molecules in polymers is the creation of gaps in the polymer matrix sufficiently large to accommodate the molecules. These gaps are formed as part of the random local polymer dynamics, which makes the diffusion of molecules in polymers a thermally activated process.

The diffusion coefficients of small molecules in polymers have been extensively studied in the literature, due to their interest for applications such as selective membranes. For instance, Barrer carried out a systematic study of the diffusion coefficient of several gases including hydrogen, nitrogen, helium, argon, oxygen, and carbon dioxide in various polymers, and established experimental correlation between the activation energy and the size of the molecules [25]. These molecules are typically referred to as small molecules due to their generally weak interaction with the polymer chains.

In the case of water, the diffusion coefficients experimentally determined are smaller than for these weakly reacting gases. In the case of Kapton polyimide, the measured diffusion coefficient in the diluted limit is 1.7×10^{-9} cm^2/s at 60 °C, with an activation energy of 5.4 kcal/g mol, while for the case of PAN the measured diffusion coefficient is 5×10^{-10} cm^2/s at 45 °C [26, 27]. Both cases are strongly concentration-dependent. Similar values have been determined for the diffusion of water in epoxy films [28].

If we consider a film of thickness L, we can define the characteristic time for diffusion t_d as:

$$t_d = \frac{L^2}{D} \tag{3.79}$$

Given a diffusion coefficient of 1×10^{-9} cm^2/s, the characteristic time to diffuse through a 1 μm thick film would be 10 s.

3.3.5 Transport in Presence of Reversible Adsorption/Desorption

The final case that will be considered in this section is transport in presence of adsorption. It is well known that the presence of adsorption processes can slow the transport of species through porous materials, a property that is at the core of applications such as gas chromatography. In the context of this book, the characteristic times for adsorption/desorption can have a large impact on the transients during the reactive transport of species [29].

To illustrate the impact of adsorption/desorption in the kinetics of transport within high surface area and nanostructured materials, we are going to consider a simple Langmuir adsorption model: if Θ represents the fractional surface coverage of an adsorbed molecule on the surface, where $\Theta = 1$ corresponds to full monolayer adsorption, the local surface kinetics can be expressed as a balance between adsorption, given by an adsorption rate $r_{\text{ads}}(\Theta, n)$, where n is the local volumetric density, and a desorption rate $r_{\text{des}}(\Theta)$:

$$\frac{d\Theta}{dt} = r_{\text{ads}}(\Theta, n) - r_{\text{des}}(\Theta) \tag{3.80}$$

If we add this model to a continuum description of the particle transport, we obtain the equation:

$$\frac{\partial n}{\partial t} - D\nabla^2 n = -\frac{\bar{s}}{s_0}\left(r_{\text{ads}}(\Theta, n) - r_{\text{des}}(\Theta)\right) = -\frac{\bar{s}}{s_0}\frac{d\Theta}{dt} \tag{3.81}$$

If the rates for adsorption and desorption are much faster than the transport rate, we can assume that coverage and local density are in local equilibrium. Consequently:

$$\frac{d\Theta}{dt} = \frac{d\Theta}{dn}\frac{\partial n}{\partial t} \tag{3.82}$$

Which means that Eq. 3.81 reduces to:

$$\frac{\partial n}{\partial t}\left(1 + \frac{\bar{s}}{s_0}\frac{d\Theta}{dn}\right) - D\nabla^2 n = 0 \tag{3.83}$$

We can rewrite Eq. 3.83 as:

$$\frac{\partial n}{\partial t} - D_{\text{eff}}\nabla^2 n = 0 \tag{3.84}$$

where:

$$D_{\text{eff}} = \frac{D}{\left(1 + \frac{\bar{s}}{s_0}\frac{d\Theta}{dn}\right)} \tag{3.85}$$

We see that the net effect of adsorption is to slow down the transport of the process by a factor that depends on the adsorption dynamics. In the limit of low fractional coverage, we have that:

$$\frac{d\Theta}{dn} \approx \frac{\Theta_0}{n_0} \tag{3.86}$$

so that:

$$D_{\text{eff}} = \frac{D}{\left(1 + \frac{\bar{s}}{s_0}\frac{\Theta_0}{n_0}\right)} \tag{3.87}$$

From this simple equation, we can estimate the impact of adsorption/desorption on transport kinetics as a function of the precursor pressure $p_0 = n_0 k_B T$ and its equilibrium adsorption coverage Θ_0.

3.4 Summary

In this chapter we have outlined the fundamentals and main approaches used to model transport in nanostructured or high surface area materials: ballistic transport models, kinetic Monte Carlo simulation, Markov chain formalism, and diffusivity-based transport models. These models will be applied in Chap. 4 to understand the main factors controlling thin film growth in high surface area materials using vapor phase thin film deposition techniques.

References

1. P. Clausing, Annalen der Physik **12**(8), 961 (1932)
2. M. Knudsen, Annalen der Physik **28**(1), 75 (1908)
3. M. Knudsen, Annalen der Physik **333**(1), 75 (1909)
4. P. Clausing, Annalen der Physik **7**(5), 569 (1930)
5. F.C. Hurlbut, J. Appl. Phys. **28**(8), 844 (1957)
6. D.H. Davis, J. Appl. Phys. **31**(7), 1169 (1960)
7. H.C. Wulu, K.C. Saraswat, J.P. Mcvittie, J. Electrochem. Soc. **138**(6), 1831 (1991)
8. A. Yanguas-Gil, J.W. Elam, Theoret. Chem. Acc. **133**(4), 1465 (2014)
9. W. Kast, C.R. Hohenthanner, Int. J. Heat Mass Transf. **43**(5), 807 (2000)
10. S.V. Sotirchos, AICHE J. **37**(9), 1365 (1991)
11. G.B. Raupp, T.S. Cale, Chem. Mater. **1**, 207 (1989)
12. S. Chatterjee, C.M. McConica, J. Electrochem. Soc. **137**, 328 (1990)
13. A. Yanguas-Gil, J.W. Elam, ECS Trans. **41**(2), 169 (2011)
14. A. Yanguas-Gil, J.W. Elam, Chem. Vap. Deposition **18**(1–3), 46 (2012)
15. S. Russ, S. Zschiegner, A. Bunde, J. Karger, Phys. Rev. E **72**(3), 030101 (2005)
16. W. Steckelmacher, Rep. Prog. Phys. **49**(10), 1083 (1986)
17. N. Wakao, J.M. Smith, Chem. Eng. Sci. **17**(11), 825 (1962)
18. L. Shen, Z.X. Chen, Chem. Eng. Sci. **62**(14), 3748 (2007)
19. J.W. Evans, M.H. Abbasi, A. Sarin, J. Chem. Phys. **72**(5), 2967 (1980)
20. W.G. Pollard, R.D. Present, Phys. Rev. **73**(7), 762 (1948)
21. J. Karger, M. Petzold, H. Pfeifer, S. Ernst, J. Weitkamp, J. Catal. **136**(2), 283 (1992)
22. P.A. Fedders, Phys. Rev. B **17**(1), 40 (1978)
23. J. Karger, Phys. Rev. A **45**(6), 4173 (1992)
24. K. Hahn, J. Karger, J. Phys. Chem. B **102**(30), 5766 (1998)
25. R.M. Barrer, J. Phys. Chem. **61**(2), 178 (1957)
26. D.K. Yang, W.J. Koros, H.B. Hopfenberg, V.T. Stannett, J. Appl. Polym. Sci. **30**(3), 1035 (1985)
27. V.T. Stannett, G.R. Ranade, W.J. Koros, J. Membr. Sci. **10**(2–3), 219 (1982)
28. M.G. McMaster, D.S. Soane, IEEE Trans. Compon. Hybrids Manuf. Technol. **12**(3), 373 (1989)
29. M.K. Gobbert, S.G. Webster, T.S. Cale, J. Electrochem. Soc. **149**(8), G461 (2002)

Chapter 4
Thin Film Growth in Nanostructured Materials

In this chapter we focus on the impact that reactive transport and surface kinetics have on the coating of nanostructured and high surface area materials. We cover a series of approximations, including line of sight, constant reaction probability, more complex non-self limited kinetics, and self-limited interactions typical from atomic layer deposition, and their impact on the ability to infiltrate and homogeneously coat high surface area materials and high aspect ratio features.

4.1 PVD and Early Line of Sight Approximations

Line of sight processes are not well-suited to coat high surface area materials, but physical vapor deposition techniques provide an efficient approach to material synthesis and are free of the complexities derived from handling chemical precursors of CVD and ALD. For that reason, they were extensively used in application domains such as semiconductor processing, and their ability and limitations to homogeneously coat nanostructured substrates are well understood. For a summary of the different approaches to increase the conformality of PVD and line of sight methods, we refer the reader to Chap. 2.

As shown in Chap. 3, if $\phi(\mathbf{x})$ represents the local incident flux to the surface, the growth rate in a line of sight process will be given simply by:

$$\mathrm{GR}(\mathbf{x}) = \frac{M_m}{\rho}\beta(\mathbf{x})\phi(\mathbf{x}) \tag{4.1}$$

The key problem in line of sight processes is therefore determining how the local incident flux changes with $\mathbf{x}$.

One of the earliest approaches to model the coating of non-planar substrates by a line-of-sight deposition process was the work by Blech [1] who focused on the

A. Yanguas-Gil, *Growth and Transport in Nanostructured Materials*,
SpringerBriefs in Materials, DOI 10.1007/978-3-319-24672-7_4

problem of coating steps using evaporation methods. Blech's model is based on the following assumptions:

1. The mean free path of atoms is larger than the distance between the evaporation source and the substrate.
2. The source-substrate distance is much larger than the step height.
3. The deposition rate is proportional to $\cos\theta_0/r^2$, where r is the source-to-substrate distance and θ_0 is the angle between the normal to the substrate and the vapor stream direction.
4. The growth takes place in the direction of the vapor stream. Rather than defining a local normal vector as is done in more advanced models in the literature, Blech defined a local velocity that was independent on the rest of the points of the surface and that was oriented towards the source. More details on surface advancement techniques will be extended in Chap. 5.
5. The effect of roughening at large angles of incidence is not considered. The formation of columns at high incidence angles in physical vapor deposition is well established and it is the basis for the deposition of sculptural films with complex 3D morphologies.
6. The sticking probability is assumed to be 1.0.

Under these assumptions, the growth rate is determined overall by the angular distribution of incident atoms, and locally by the subset of angles that provide line of sight with the substrate. Tisone and Bindell provide a detailed analysis on the application of this method to model the growth on nanostructured substrates from extended sputtering sources [2, 3].

4.1.1 *Directed Flow*

If we have a directed or fully collimated flow, the growth rate can take only two possible values depending on the existence of an unobstructed line-of-sight path between a particular point of the surface and the source of particles. This is shown in Fig. 4.1 for two cases: a mask with a 45 degrees incidence angle, and a trench with a vertical fully collimated flow.

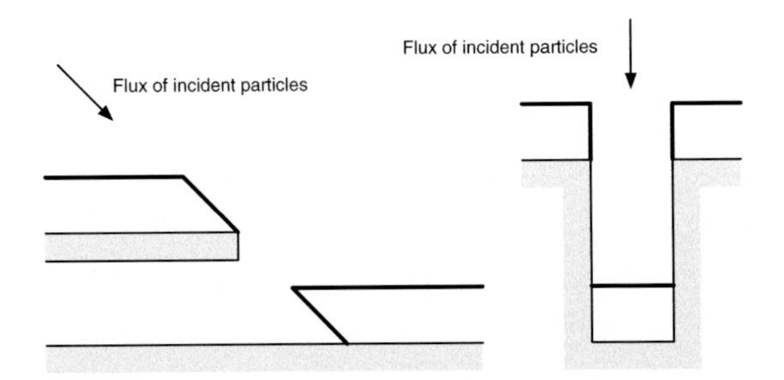

Fig. 4.1 Ideal profile evolution for collimated incident flow

Figure 4.1 also emphasizes the importance of controlling the angular distribution function of molecules. A highly collimated normal flow ideally allows to efficiently deposit at the bottom of high aspect ratio features, since it avoids sidewall collisions. Strategies aimed at improving the conformality of PVD methods have focused on devising means for achieving this, such as through collimated PVD or the use of ionized PVD, which achieves a highly directional flow through the ionization of the sputtered material.

4.1.2 Isotropic Case

If an isotropic distribution of incident particles is instead considered, the growth rate at every point will be determined by the solid angle accessible from the gas phase at that point of the surface. In the case of a structure with translational symmetry in one direction, such as a step, this can be done by tracking the two angles φ^- and φ^+ with respect to the normal describing the local shadowing of a point on the surface. This was the approached followed by Blech.

However, in the case of isotropic deposition one can take advantage of the concept of view factor introduced in Chap. 3 to determine the local flux to the surface. Here we are going to focus on two examples: the thickness on the sidewalls in a rectangular trench and in a circular via (Fig. 4.2).

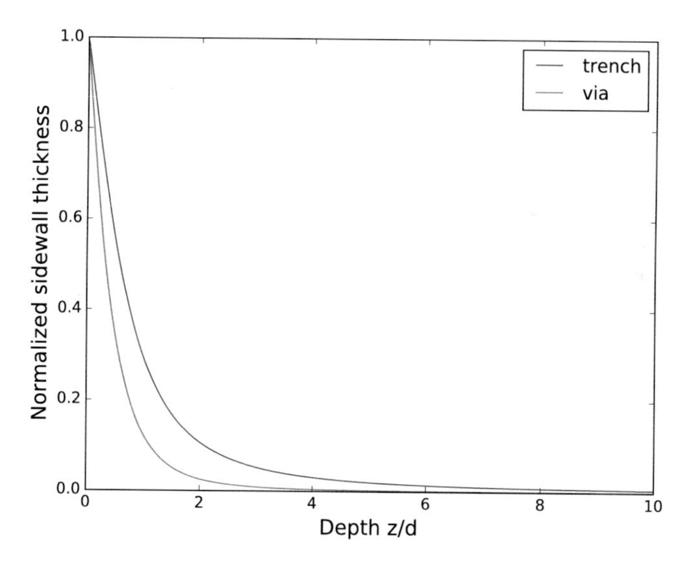

Fig. 4.2 Sidewall thickness as a function of depth for a circular via and a rectangular trench for a line of sight process with isotropic source

4.1.2.1 Rectangular Trenches

Given a trench of diameter d, the probability that a particle arriving from the top hits the sidewalls at a depth z is given by Eq. 3.28, reproduced here as a function of z and d:

$$\mathrm{GR}(z) = \mathrm{GR}_0 \left[1 - \frac{z}{\left(d^2 + z^2\right)^{1/2}} \right] \tag{4.2}$$

where GR_0 is the growth rate at the top of the trench.

4.1.2.2 Circular Pores or Vias

For a circular via, the resulting expression is given by:

$$\mathrm{GR}(z) = \mathrm{GR}_0 \left[\frac{2}{d} \left(\frac{2z^2 + d^2}{2\sqrt{z^2 + d^2}} - z \right) \right] \tag{4.3}$$

Figure 4.2 shows the reduction in film thickness on the sidewalls as a function of depth for a rectangular trench and a circular via. It is clear that achieving good step coverages in circular vias is much harder than in trenches, and that for a line of sight process achieving good step coverage in aspect ratios greater than 1 becomes challenging.

The values provided in Eqs. 4.2 and 4.3 correspond to the initial deposition step. As growth proceeds, overhangs will develop on the top of the trench until inevitably pinch-off occurs. In Chap. 2 several strategies were described to mitigate this problem. Approaches used in the literature include taking advantage of surface mobility to enhance surface transport, the modulation of the angular distribution function, and the incorporation of simultaneous directed etching by ion bombardment.

4.2 Constant Sticking Probability

An important generalization of line-of-sight models is the assumption that incident species react with the surface with a constant sticking probability β. While in many instances, particularly in CVD, the sticking probability depends on the state of the surface and can be also surface-flux dependent, the constant sticking probability model is a good assumption whenever the precursor is on the transport-limited regime and the growth rate is therefore linearly dependent with the precursor pressure.

The fact that $\beta < 1$ implies that molecules can undergo multiple collisions with the surface, which help them penetrate deeper into a nanostructured material. From an statistical point of view, the probability $P(n)$ that a molecule reacts after n collisions will be given by:

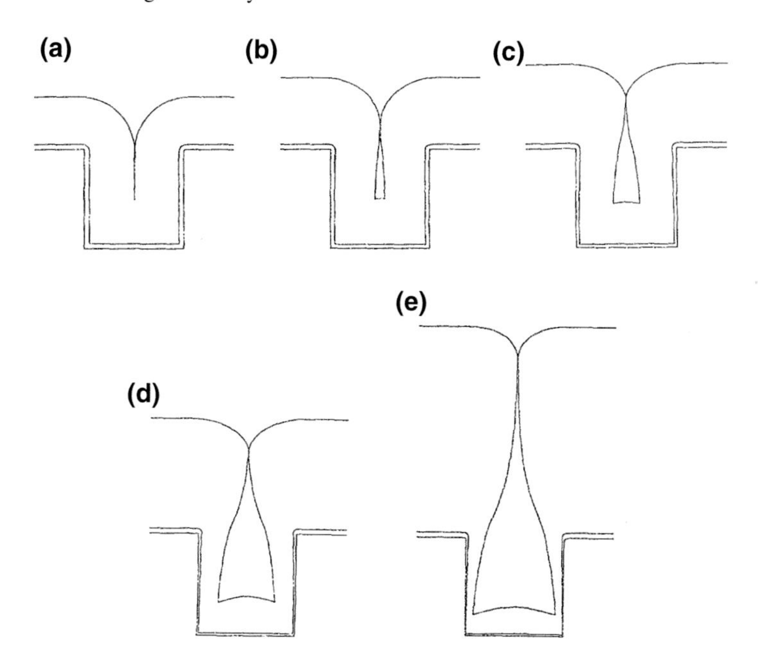

Fig. 4.3 Growth profiles on a trench simulated using a ballistic model in a rectangular trench for different values of the sticking probability: 0.01 (**a**), 0.1 (**b**), 0.2 (**c**), 0.5 (**d**), and 1 (**e**). Reproduced from Ref. [4] with permission

$$P(n) = (1 - \beta)^{n-1}\beta \tag{4.4}$$

and the average number of collisions N is simply:

$$N = \frac{1}{\beta} \tag{4.5}$$

Consequently, the value of β has a great impact on the process' ability to coat high aspect ratio features. One example is shown in Fig. 4.3, where simulated profiles of growth on rectangular trenches are presented for different values of the sticking probability [4]. Figure 4.3 is a great example of how the ability to completely fill even moderate aspect ratio rectangular trenches can be compromised when the sticking probability is too high. In this particular case, above $\beta = 0.01$ pinch off occurs before the trench is fully filled, leaving a gap inside the feature.

Another example is the opal structure shown in Fig. 4.4. Silica and polystyrene opals are used as template for the formation of photonic structures such as inverse opals or inverse eggshell structures. In addition to self-limited techniques like ALD, examples in the literature of opal formation based on CVD processes include Si and Ge from SiH_4 and Ge_2H_6 and HfB_2 from $Hf(BH_4)_4$ [5–7].

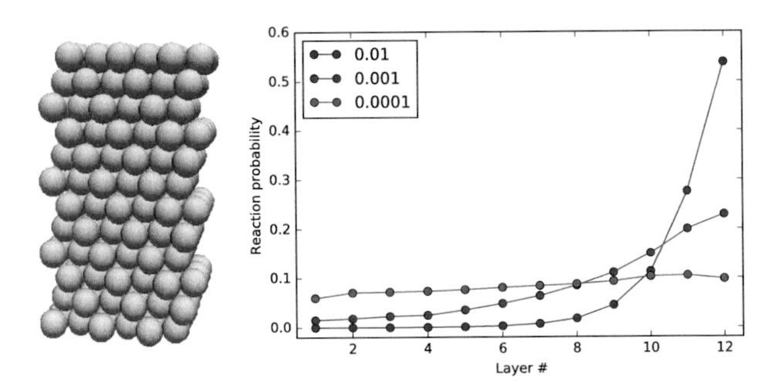

Fig. 4.4 Opal structure formed by 12 layers of spheres arranged in a fcc 111 packing (*left*) and the probability that an incident molecule reacts at each of the layers for three values of the sticking probability: 0.01, 0.001, 0.0001. For $\beta = 0.0001$, the coating is almost conformal

We can therefore use a Monte Carlo model like the ones described in Chap. 3 to model the trajectories of incident precursor molecules inside an opal film and predict where an incident precursor molecule will react within the structure. The probability that a particle reacts at the different layers of the opal structure is shown in Fig. 4.4 for three values of the sticking probability: 0.01, 0.001, and 0.0001. The figure shows that processes characterized by an sticking probability of 0.0001 can conformally coat a 12 layer opal structure.

4.2.1 Diffusion-Based Model

If we model the transport inside the high surface area material as a diffusive process, as described in Chap. 3, we can determine the growth rate as a function of depth by solving a simple diffusion-reaction equation.

Using the diffusion equation developed in Chap. 3, we have that the reactive transport of precursor molecules inside a high surface area material can be modeled with the equation:

$$D\frac{d^2n}{dz^2} = \bar{s}\beta(z)\phi(z) = \bar{s}\frac{1}{4}\bar{v}\beta(z)n \tag{4.6}$$

Here n is the precursor density, which is related to the precursor pressure p as: $p = nk_BT$, D is a diffusivity whose value will depend on the regime (Knudsen, transitional, or bulk diffusion) and the microstructure of the material, $\bar{s}$ represents the specific surface area, that is, the surface area per unit volume, and the product:

$$\phi(z) = \frac{1}{4}\bar{v}n(z) \tag{4.7}$$

represents the surface flux at every depth of the feature.

We can normalize Eq. 4.6 in terms of the thickness of the nanostructured material L, resulting on an equation which depends on a single parameter:

$$\frac{d^2n}{d\xi^2} = \frac{\bar{s}L^2\bar{v}\beta}{4D}n = h_T^2\, n \tag{4.8}$$

The parameter h_T is the so-called *Thiele modulus*, which is simply the ratio between the reaction rate and the transport rate, given by:

$$h_T^2 = \frac{\frac{1}{4}\bar{s}\bar{v}\beta}{D/L^2} \tag{4.9}$$

Based on the local density, the growth rate at every point in the feature will be simply given by:

$$\mathrm{GR} = \frac{M_m}{\rho}\beta\frac{1}{4}\bar{v}n \tag{4.10}$$

Consequently, the conformality of the process will be determined by the gradient in precursor concentration $n(z)$ inside the high surface area material.

Equation 4.6 constitutes the basis of the analysis carried out in this chapter to understand the impact of surface reactivity on the conformality of a given process. The key advantage with respect to ballistic approaches highlighted in Figs. 4.3 and 4.4 is that by solving exactly Eq. 4.8 or similar equations, we can extract the dependence of conformality with the surface kinetics of the growth process.

4.2.2 *Impact of Reaction Probability on Film Conformality*

To investigate the impact that the constant reaction probability β and the structure of a nanostructured or high surface area material have on conformality we can simply solve Eq. 4.8 subject to the following boundary conditions. At the top ($z = 0$) we will impose a constant value of precursor density:

$$n(0) = n_0 \tag{4.11}$$

While at the bottom the diffusive flux must match the losses at the bottom surface:

$$-D\left.\frac{dn}{dz}\right|_{z=L} = \frac{1}{4}\bar{v}\beta n(L) \tag{4.12}$$

or

$$-\left.\frac{dn}{d\xi}\right|_{\xi=1} = \frac{1}{\bar{s}L}h_T^2 n(1) = \frac{1}{s_e}h_T^2 n(1) \tag{4.13}$$

where s_e is simply the surface area enhancement, defined as the total surface area per substrate surface area,

$$s_e = \bar{s}L \tag{4.14}$$

Defining the step coverage SC as the thickness at the bottom divided by the thickness at the top, we have that:

$$\mathrm{SC} = \frac{2s_e e^{-h_T}}{(s_e + h_T) + (s_e - h_T)e^{-2h_T}} \tag{4.15}$$

If the surface area enhancement of the material is high enough, this expression reduces to:

$$\mathrm{SC} = \frac{2e^{-h_T}}{1 + e^{-2h_T}} \tag{4.16}$$

Therefore, if $s_e \gg h_T$, the ability of a given process to coat a high surface area material depends solely on the Thiele modulus.

In Fig. 4.5 the expected step coverage of a process is shown as a function of the Thiele modulus for different values of surface area enhancement s_e.

The transition between a conformal and a non-conformal process is determined by the condition $h_T = 1$. If

$$h_T < 1 \tag{4.17}$$

the process will be able to coat conformally the high surface area material.

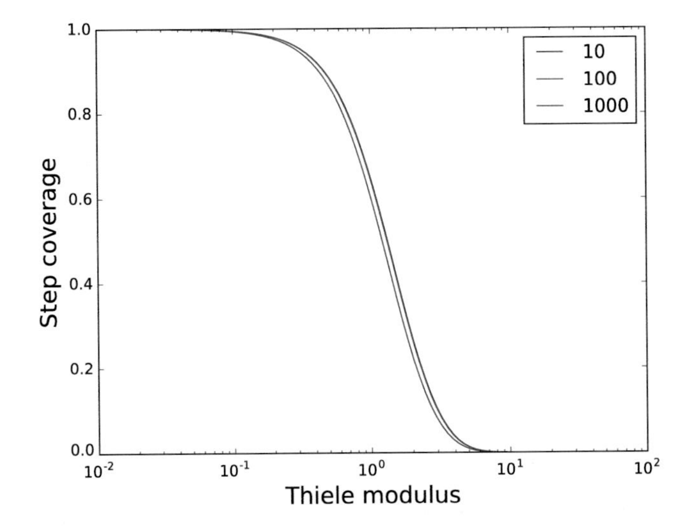

Fig. 4.5 Dependence of the step coverage on the Thiele modulus for different values of the surface area enhancement s_e (total surface area per unit surface area)

So what factors affect the value of h_T? It is clear from Eq. 4.9 that the sticking probability, the thickness of the material, and the surface area all tend to increase the Thiele modulus and therefore reduce the conformality of the deposition process. In contrast, fast transport through the pores allows for a more even coating of the nanostructured material.

Next we particularize these results for two different cases: reactive transport within circular pore or via and the reactive transport inside a random porous material.

4.2.2.1 Circular Pores and Vias

If we assume that the pore diameter is smaller than the mean free path of the precursor molecules, then transport is driven by Knudsen diffusion. In this case:

$$D = \frac{1}{3}\bar{v}d \tag{4.18}$$

The surface area per unit volume becomes simply the inverse of the pore diameter:

$$\bar{s} = \frac{4}{d} \tag{4.19}$$

And the expression of the Thiele modulus reduces to:

$$h_T^2 = \frac{3L^2}{d^2}\beta_0 = 3(AR)^2\beta_0 \tag{4.20}$$

or

$$h_T = \sqrt{3\beta_0}(AR) \tag{4.21}$$

where we have defined the aspect ratio as the depth to diameter ratio. This allows us to establish a criterion for achieving a conformal growth on high aspect ratio vias. Conformal deposition will result whenever the condition:

$$\beta_0 < \frac{1}{3(AR)^2} \tag{4.22}$$

is satisfied.

This equation codifies what was advanced in Fig. 4.3: by choosing processes with a low reaction probability we can improve the step coverage of a CVD process. In Fig. 4.6 we show a set of curves in which we represent the reaction probability required to achieve a certain step coverage as a function of the aspect ratio for circular pores. This figure shows the results for aspect ratio 10 or higher since, as shown in Chap. 3, the expression of the diffusion coefficient used to model the transport within the pores assumes that pores are infinitely long. For shorter aspect ratios,

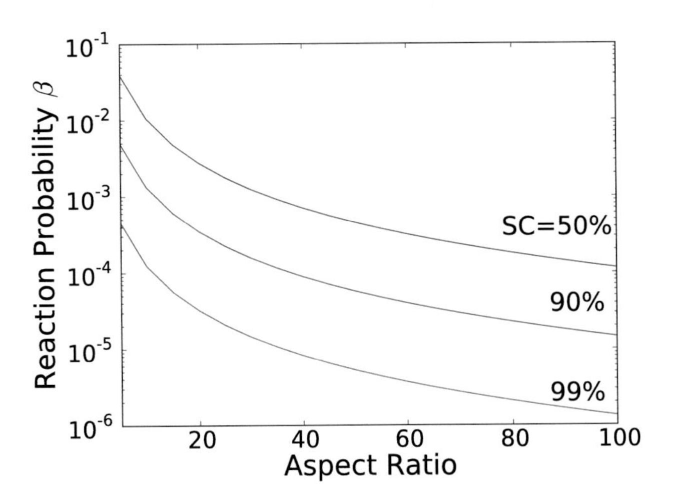

Fig. 4.6 Reaction probabilities required to achieve target step coverages in a circular via as a function of the aspect-ratio

modifications to the diffusion coefficient to take into account entry effects must be considered [8].

Finally, this result also confirms a well-known result from the CVD literature: when the transport takes place under the Knudsen regime, the ability of the process to grow conformally inside of a high surface area material depends on its specific surface area, in this case the aspect ratio of the pore, and not on the actual dimension of the feature.

4.2.2.2 Porous Materials

In the general case of a porous material, we can use Eq. 3.62 in which the diffusivity is expressed in terms of the Knudsen diffusion coefficient for the average pore size d, the porosity ε of the material and the tortuosity τ:

$$D = \frac{1}{3}\bar{v}d\frac{\varepsilon}{\tau} \tag{4.23}$$

With that, the Thiele modulus now becomes:

$$h_T^2 = \frac{\frac{3}{4}\bar{s}\beta L^2}{d}\frac{\tau}{\varepsilon} \tag{4.24}$$

Larger tortuosity values negatively affect the ability to conformally coat a porous material.

4.2.2.3 Conformality in the Transitional Regime

So far in the examples above we have assumed that the mean pore diameter is smaller than the mean free path, so that transport is dominated by Knudsen diffusion. If pressure or the pore diameter is large enough, the molecular component can no longer be neglected, and we have to consider diffusive transport in the transitional regime.

As shown in Chap. 3, the Bosanquet approximation is a good approximation for this transitional regime:

$$\frac{1}{D} = \frac{1}{D_K} + \frac{1}{D_m} \tag{4.25}$$

where D is the diffusion coefficient across the whole pressure range, D_K is the Knudsen diffusivity, and D_m is the molecular diffusivity. If for simplicity, we assume that our precursor molecule is diluted by a carrier gas, then we can approximate the molecular diffusion coefficient such that:

$$D_m = \frac{A}{p} \tag{4.26}$$

This creates a pressure dependence in the effective diffusion coefficient, which can be expressed as follows:

$$D = \frac{D_K}{1 + \frac{D_K p}{A}} \tag{4.27}$$

The presence of a pressure-dependent transport coefficient makes the step coverage of a process change with pressure. This was for instance identified in chemical vapor change infiltration experiments, where the characteristic pore sizes are large enough for the transport to be influenced by gas phase collisions.

4.3 Pressure-Dependent Kinetics

The constant reaction probability approximation of the previous section is a good approximation for PVD and some CVD processes. However, it is unable to capture the complexity of many other systems. In this section we will introduce a general approach to understand the impact of surface kinetics on conformality.

Let us consider a process whose growth rate depends on the pressure of the different species:

$$\mathrm{GR} = \mathrm{GR}(p_i, T) \tag{4.28}$$

where p_i is the pressure of i, species. Strictly speaking, the growth rate will depend on the local fluxes ϕ_i to the surface, but in CVD the presence of local thermodynamic equilibrium is a good approximation except perhaps under ultrahigh vacuum conditions, and we can assume that pressure and surface flux are related through:

$$\phi_i = \frac{1}{4}\bar{v}\frac{p_i}{k_B T}. \tag{4.29}$$

We can use Eq. 4.28 to define the conformality of a process in terms of the pressure depletion due to its consumption through surface reactions inside the high surface area material, so that the growth rate at the bottom of the porous material is simply given by:

$$\mathrm{GR}_{\mathrm{bottom}} = \mathrm{GR}(p_i - \Delta p_i, T) \tag{4.30}$$

The step coverage SC can be then expressed as:

$$\mathrm{SC} = \frac{\mathrm{GR}_{\mathrm{bottom}}}{\mathrm{GR}_{\mathrm{top}}} = \frac{\mathrm{GR}(p_i - \Delta p_i, T)}{\mathrm{GR}(p_i, T)} \tag{4.31}$$

where Δp_i is the pressure drop due to the consumption of species i inside the material.

Let us now focus on a situation where the step coverage is fair: from Eq. 4.31 this implies that $\mathrm{GR}(p_i - \Delta p_i, T)$ will be at least of the order of $\mathrm{GR}(p_i, T)$. This allows us to carry out a first order expansion of the growth rate centered around $\mathrm{GR}(p_i, T)$ so that:

$$\mathrm{GR}(p_i - \Delta p_i, T) \approx \mathrm{GR}(p_i, T) + \sum_i \frac{\partial \mathrm{GR}}{\partial p_i}\Delta p_i \tag{4.32}$$

This yields the following expression for the step coverage:

$$\mathrm{SC} = 1 + \frac{1}{\mathrm{GR}_0}\sum_i \frac{\partial \mathrm{GR}}{\partial p_i}\Delta p_i \tag{4.33}$$

where $\mathrm{GR}_0 = \mathrm{GR}(p_i, T)$.

From Eq. 4.33 we see that the step coverage will be the product of two factors: the precursor depletion Δp_i and the dependence of growth rate with the pressure of species i.

If all Δp_i are less than zero, that is, all the relevant species are being consumed as part of the growth process, we can take the absolute value of the pressure drop $|\Delta p_i|$ and instead use a negative sign in Eq. 4.33 so that:

$$\mathrm{SC} = 1 - \frac{1}{\mathrm{GR}_0}\sum_i \frac{\partial \mathrm{GR}}{\partial p_i}|\Delta p_i| \tag{4.34}$$

In the remaining of this section we will explore the different implications of Eq. 4.33 and the way in which it can be applied to rationalize the impact that surface kinetics has on the ability of a process to coat nanostructured or high surface materials. This will also allow us to design strategies to improve the conformality of existing processes, as shown later in this chapter.

4.3.1 Conformal Zone for Single-Source Precursors

In this section we particularize Eq. 4.33 to processes in which growth is driven by a single species. This section will follow closely the discussion presented in Ref. [9].

In a process controlled by only a single species:

$$\mathrm{SC} = 1 - \frac{1}{\mathrm{GR}(p,T)} \frac{\partial \mathrm{GR}(p,T)}{\partial p} \Delta p \tag{4.35}$$

Our goal is to use Eq. 4.35 and our knowledge of GR (p, T) to determine a condition similar to Eq. 4.17 that provides a requirement to conformally coat a high surface area material.

The key is how to obtain an expression for the precursor depletion inside the high surface area material. For that, we see that if the step coverage is good, the variation of GR inside the high surface area is going to be small. Therefore, we can estimate Δp assuming a constant growth rate, so that the reactive transport of the precursor inside the nanostructured feature is driven by the equation:

$$D \frac{d^2 n}{dz^2} = \bar{s}\mathrm{GR} \tag{4.36}$$

subject to the boundary conditions:

$$n(0) = n_0 = \frac{p}{k_B T} \tag{4.37}$$

and

$$-D \left. \frac{dn}{dz} \right|_{z=L} = \mathrm{GR} \tag{4.38}$$

Note that, in order to keep the dimensions consistent, the growth rate GR is in units of atoms (or moles) per unit area. This results in an expression for Δp given by:

$$\Delta p = \frac{\Delta n}{kT} = \frac{\bar{s}\,\mathrm{GR} L^2}{k_B T D} \left(\frac{1}{2} + \frac{1}{\bar{s} L} \right) \tag{4.39}$$

And combining Eqs. 4.35 and 4.39 we have that:

$$\mathrm{SC} = 1 - \frac{\partial \mathrm{GR}(p,T)}{\partial p} \frac{\bar{s} L^2}{k_B T D} \left(\frac{1}{2} + \frac{1}{s_e} \right) \tag{4.40}$$

where we have also made use of the definition of the surface area enhancement $s_e = \bar{s} L$.

There are several conclusions that we can extract from Eq. 4.40:

1. The step coverage does not depend on the absolute growth rate or the precursor pressure, but rather on the dependence of the growth rate with pressure. A weak dependence will ensure a good step coverage, something that has long been recognized in CVD and CVI.
2. The condition of having a small sticking probability in order to achieve conformal growth can be considered a particular case of Eq. 4.40. When the sticking probability is constant over a large range of pressures, the derivative of the growth rate with pressure is proportional to the sticking probability:

$$\frac{\partial \mathrm{GR}(p, T)}{\partial p} \sim \beta \tag{4.41}$$

3. The morphology of the nanostructured substrate affects the step coverage through three different parameters: the specific surface area $\bar{s}$, the thickness of the substrate L and the diffusivity of the material D.
4. The rest of parameters being equal, pressure can have a detrimental effect on the step coverage due to the decrease of the diffusion coefficient with pressure at the transitional regime and beyond: Knudsen diffusion will lead to a higher, pressure-independent diffusivity, while the molecular diffusivity decreases with pressure as $1/p$.

If we select a target step coverage SC_0, Eq. 4.40 implies that, for a given CVD process, there is only a subset of the p, T parameter space in which that target can be achieved for a certain substrate. This subset is defined by the condition:

$$\frac{\partial \mathrm{GR}(p, T)}{\partial p} \frac{\bar{s}L^2}{k_B T D} \left(\frac{1}{2} + \frac{1}{s_e} \right) \leq 1 - \mathrm{SC}_0 \tag{4.42}$$

If we assume a diffusivity given by Eq. 4.23, then the boundary for conformality is given by:

$$\frac{\partial \mathrm{GR}(p, T)}{\partial p} \frac{3\bar{s}L^2}{k_B T \bar{v} d} \frac{\tau}{\varepsilon} \left(\frac{1}{2} + \frac{1}{s_e} \right) \leq 1 - \mathrm{SC}_0 \tag{4.43}$$

Here, $\mathrm{GR}(p, T)$, the tortuosity τ, the accessible porosity ε, the average pore size d and the thickness L of the high surface area material are all known variables. Since GR is expressed in terms of atoms per unit area and time, we need to know the density of the film and its stoichiometry, or get the direct growth rate using techniques such as Rutherford Backscattering Spectroscopy, which can directly provide the atoms per unit area in a film.

4.3.1.1 First Order Langmuir Kinetics

The dependence of growth rate with pressure in many CVD processes follows a similar trend: at very low pressures the growth rate is proportional to precursor pressure and as the pressure increases the growth rate rolls over until reaching a saturation value, at which point the growth rate becomes essentially independent of pressure. This corresponds to the reaction limited regime. Its weak pressure dependence makes it the ideal regime for the coating of high surface area materials, since the growth rate is less sensitive to precursor depletion inside the nanostructured material.

One of the simplest models that is able to reproduce the dependence with temperature and pressure found in many single-source precursor systems is the first order Langmuir kinetics. This model assumes that the overall reaction takes place through a surface intermediate:

$$\mathrm{A_{(g)}} \rightleftharpoons \mathrm{A_{(s)}} \rightarrow \mathrm{M_{(s)}} + \mathrm{Byproducts} \tag{4.44}$$

This model results on the following dependence of the growth rate with p and T:

$$\mathrm{GR}(p, T) = \frac{K_1(T)p}{1 + K_2(T)p} \tag{4.45}$$

Substituting Eq. 4.45 in Eq. 4.40, we obtain that:

$$\mathrm{SC} = 1 - \frac{K_1(T)}{(1 + K_2(T)p)^2} \frac{\bar{s}L^2}{k_B T D} \left(\frac{1}{2} + \frac{1}{s_e} \right) \tag{4.46}$$

If we set a threshold value of step coverage SC_0, the subset of experimental conditions in the (p, T) space that will be able to meet this requirement will be constrained by the following condition:

$$\mathrm{SC}_0 \geq 1 - \frac{K_1(T)}{(1 + K_2(T)p)^2} \frac{\bar{s}L^2}{k_B T D} \left(\frac{1}{2} + \frac{1}{s_e} \right) \tag{4.47}$$

or

$$\frac{(1 + K_2(T)p)^2}{K_1(T)} \geq \frac{1}{1 - \mathrm{SC}_0} \frac{\bar{s}L^2}{D k_B T} \left(\frac{1}{2} + \frac{1}{s_e} \right) \tag{4.48}$$

Equation 4.48 defines the subset in the (p, T) parameter space for which the CVD process is able to conformally coat the high surface area or nanostructured material.

If the surface area enhancement s_e of the material and the pressure are high enough so that $K_2(T)p \gg 1$, this equation reduces to:

$$p^2 \geq \frac{1}{1 - \mathrm{SC}_0} \frac{K_1(T)}{K_2(T)^2} \frac{\bar{s}L^2}{2 D k_B T} \tag{4.49}$$

Pressure becomes clearly the key parameter that must be tuned to accommodate the high surface area and the thickness of a material: high surface area materials will require higher precursor pressures to maintain a weak dependence of the growth rate with pressure everywhere within the high surface area material.

On the other hand, if $K_2(T) = 0$, then the condition reduces to:

$$K_1(T) \leq (1 - \mathrm{SC}_0)\frac{2Dk_BT}{\bar{s}L^2} \tag{4.50}$$

Pressure has now disappeared from the equation, and the only way in which the conformality of the process can be tuned is by changing the temperature to adjust the value of $K_1(T)$. This is equivalent to tuning the sticking probability by controlling the temperature of the process.

4.3.1.2 Conformal Zone Diagram

Our ability to increase the pressure or reduce the temperature arbitrarily is not absolute: we are limited by factors such as the vapor pressure, and the limit imposed by gas-phase collisions which would introduce further complications on the surface kinetics. This means that depending on how stringent the conditions are, a conformal zone may simply not exist for many systems. As shown in Fig. 4.7, too low vapor pressures or too high CVD onset may substantially limit the range of aspect ratio accessible to a particular process, all the other conditions being equal.

Together with these other constraints, Eq. 4.48 provides the boundary of the conformal zone of a CVD process (Fig. 4.8).

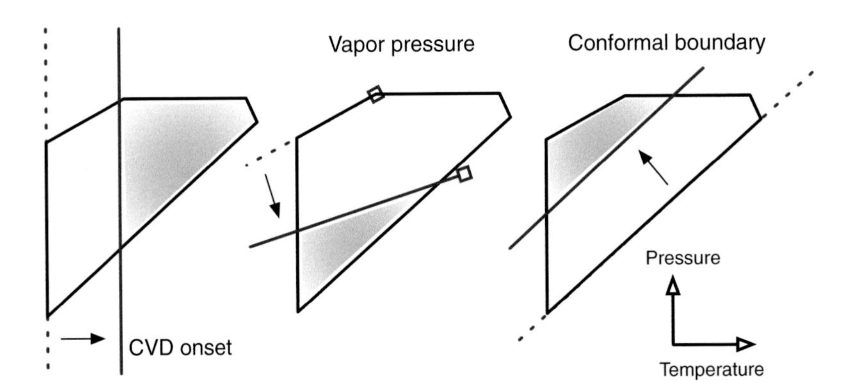

Fig. 4.7 Factors affecting the existence of a conformal region in CVD include the reaction temperature onset, the precursor vapor pressure, and the position of the conformal boundary defined by Eq. 4.40, which itself depends on surface kinetics and the nature of the nanostructured substrate

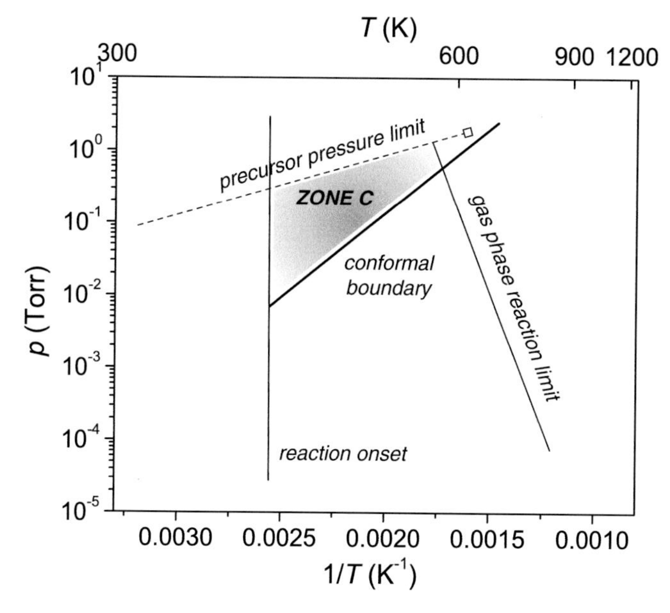

Fig. 4.8 Conformal zone diagram for a low pressure CVD process controlled by a single precursor: the *shaded area* represents the subset of the parameter space where conformal growth can be achieved for a given substrate Reproduced from Ref [9] with permission

4.3.2 *Superconformal Processes and Conformality Enhancement*

We can extend the reasoning introduced in the previous section to processes involving more than one species. If we assume that we have two species, then Eq. 4.33 reduces to:

$$\mathrm{SC} = 1 - \frac{1}{\mathrm{GR}_0}\left(\frac{\partial \mathrm{GR}}{\partial p_1}|\Delta p_1| + \frac{\partial \mathrm{GR}}{\partial p_2}|\Delta p_2|\right) \tag{4.51}$$

In many cases this means that we now have to worry about the depletion of two different species.

However, it is also possible to conceive a process in which the second species has the opposite effect on the growth rate, so that:

$$\frac{\partial \mathrm{GR}}{\partial p_2} < 0 \tag{4.52}$$

In that case, if we take absolute values, we have that the step coverage is given by:

$$\mathrm{SC} = 1 - \frac{1}{\mathrm{GR}_0}\left(\left|\frac{\partial \mathrm{GR}}{\partial p_1}\right||\Delta p_1| - \left|\frac{\partial \mathrm{GR}}{\partial p_2}\right||\Delta p_2|\right) \tag{4.53}$$

The depletion of the second species actually helps increasing the conformality of the process. Moreover, if we have that:

$$\left|\frac{\partial \mathrm{GR}}{\partial p_2}\right| |\Delta p_2| > \left|\frac{\partial \mathrm{GR}}{\partial p_1}\right| |\Delta p_1| \tag{4.54}$$

we have that SC $>$ 1: the growth rate inside the high surface area material is actually higher than at the top of the high surface area material. Processes exhibiting this property are said to be *superconformal*.

From a mechanistic point of view, there are several ways in which this could be accomplished.

4.3.2.1 Simultaneous Growth and Etching

A first possible scenario is the presence of simultaneous growth and etching. In the context of PVD, this is accomplished by accelerating ions towards the substrate, which causes sputtering and redeposition. In the case of chemistry-based processes, there are some systems for which the simultaneous presence of growth and etching has been established. One such example is the epitaxial growth of SiC by CVD. This process is typically carried out using hydrogen as a carrier gas, and in absence of precursors, the etching of SiC by hydrogen has been reported by different authors [10].

In both cases, the presence of a simultaneous growth and etching causes a redistribution of the mass being deposited that can affect conformality. The growth rate at any point inside the high surface area material GR will be given by the net balance between the growth and etching processes. Let n the precursor density, n_e the density of the etching species, and n_r the density of any reemitted species. The growth rate can be expressed as:

$$\mathrm{GR} = s_0\beta_0\frac{1}{4}\bar{v}n - s_0\beta_e\frac{1}{4}\bar{v}_e n_e + \beta_r\frac{1}{4}\bar{v}_r n_r = g_0 - g_e + g_r \tag{4.55}$$

where β_0, β_e, and β_r are the sticking probabilities of the corresponding processes.

If we model the transport inside the porous or nanostructured material as a diffusive process, we have the following set of equations for this simple model:

$$D\frac{d^2n}{dz^2} = \beta_0\frac{1}{4}\bar{v}n \tag{4.56}$$

$$D_e\frac{d^2n_e}{dz^2} = \beta_e\frac{1}{4}\bar{v}_e n_e \tag{4.57}$$

$$D_r\frac{d^2n_r}{dz^2} = \beta_r\frac{1}{4}\bar{v}n_r - \beta_e\frac{1}{4}\bar{v}_e n_e \tag{4.58}$$

To our knowledge, no in-depth study has been carried out of processes involving simultaneous growth and etching inside nanostructured materials using these questions. As an example, if we focus on a system where the impact of redeposition can be neglected, $\beta_r = 0$, we obtain the following expression for the step coverage:

$$\mathrm{SC} = \frac{g_0\mathrm{SC}_0 - g_e\mathrm{SC}_e}{g_0 - g_e} \tag{4.59}$$

Here, the growth rate GR is expressed as the net effect between the growth process, characterized by an ideal growth rate g_0, and the etching process, characterized by an etch rate g_e. SC_0 and SC_e are the step coverage of these two separate processes, as given by Eq. 4.15. We can further simplify this expression if we define r_e as the etching to growth ratio:$r_e = g_e/g_0$:

$$\mathrm{SC} = \frac{\mathrm{SC}_0 - r_e\mathrm{SC}_e}{1 - r_e} \tag{4.60}$$

In this simple case, superconformal growth can be achieved whenever:

$$\mathrm{SC}_e < \frac{\mathrm{SC}_0 - (1 - r_e)}{r_e} \tag{4.61}$$

From an experimental point of view, superconformal deposition in presence of simultaneous growth and etching requires an etching process with poor step coverage, that is, that preferentially etches the top of the surface, with the threshold value given by the step coverage of the deposition component SC_0 and the ratio between the etch and deposition rates r_e.

4.3.2.2 Competitive Adsorption

A second approach to achieve superconformal deposition is one involving two precursors that compete for the same pool of adsorption sites on the surface.

Let us assume that the growth rate is first order in the surface coverage of each species Θ_1 and Θ_2:

$$\mathrm{GR} = k_g\Theta_1\Theta_2 \tag{4.62}$$

We can obtain the values of Θ_1 and Θ_2 from the balance between gain (adsorption) and loss (desorption and reaction) processes:

$$\alpha_1 p_1(1 - \Theta_1 - \Theta_2) - k_1\Theta_1 = k_g\Theta_1\Theta_2 \tag{4.63}$$

$$\alpha_2 p_2(1 - \Theta_1 - \Theta_2) - k_2\Theta_2 = k_g\Theta_1\Theta_2 \tag{4.64}$$

In excess of precursor 1, $\Theta_1 \gg \Theta_2$, and we can approximate the growth rate as:

$$\mathrm{GR}(p_1, p_2) = \frac{ap_1p_2}{(1 + bp_1)(1 + cp_1)} \tag{4.65}$$

where a, b, and c are three temperature-dependent coefficients that, for this simple model, are given by:

$$a = \frac{k_g \alpha_1 \alpha_2}{k_1 k_2} \tag{4.66}$$

$$b = \frac{\alpha_1}{k_1} \tag{4.67}$$

$$c = \frac{\alpha_1 (k_g + k_2)}{k_1 k_2} \tag{4.68}$$

It is clear that in excess of precursor 1, the growth rate actually decreases with p_1, so that GR $\sim 1/p_1$.

From an experimental perspective, one system that shows the same qualitative behavior is the growth of MgO from $Mg(DMDBA)_2$ and H_2O. The dependence of the growth rate with water pressure exhibits a characteristic volcano shape, with the growth rate decreasing with increasing pressure beyond a certain critical pressure value. This lead to highly conformal growth and the bottom up filling of V-shaped vias as described by Wang et al. [11, 12].

4.3.2.3 Enhancing Conformality Through Growth Inhibition

The central concept in this approach is to modify precursor-surface interaction by introducing additional processes that can help modulate the sticking probability [13].

If we consider what a constant reaction probability means in terms of precursor-surface interaction, one of the main consequences is that the growth rate is locally rate limited by the flow of species. In particular, it neglects the presence of a substantial fraction of surface intermediates on the surface. However, the sticking probability is still sensitive to certain surface parameters: for instance, if the surface density of reactive sites can somehow be reduced, the effective sticking probability will become smaller if adsorbed species cannot adsorb and diffuse over large distances.

For simplicity, let's assume a single precursor process which is characterized by a pressure-independent sticking probability β_0. If we now add a second species that blocks a fraction of the surface sites Θ_i, the sticking probability is reduced with respect to the original by a factor $1 - \Theta_i$:

$$\beta = (1 - \Theta_i)\beta_0 \tag{4.69}$$

Thereby improving the step coverage of the additional process. The challenge in this case is to identify such an species.

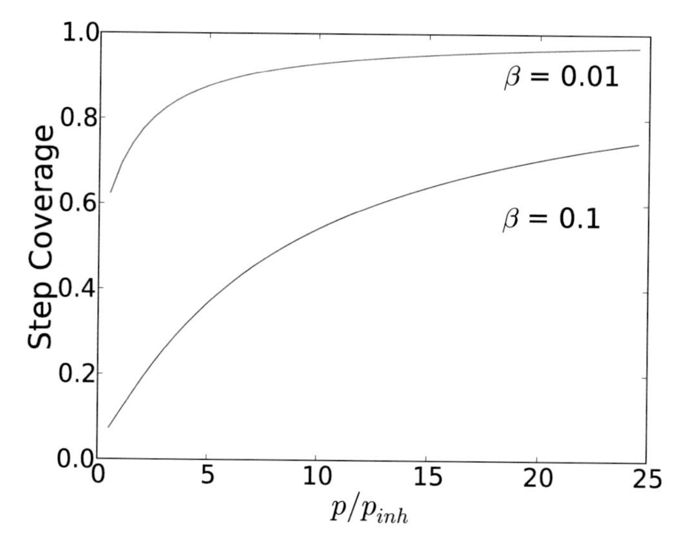

Fig. 4.9 Impact of growth inhibition in the step coverage in a 10 aspect ratio feature for two initial reaction probabilities: $\beta = 10^{-1}$ and $\beta = 10^{-2}$. Reproduced from Ref. [13] with permission

Alternatively, if a precursor undergoes dissociative adsorption:

$$AB_g \rightarrow A_{ads} + B_{ads} \tag{4.70}$$

With B being typically one or more of the precursor ligands, in many cases a balance is established between this process and the reverse, associative desorption:

$$AB_g \rightleftharpoons A_{ads} + B_{ads} \tag{4.71}$$

In this case, the presence of an additional partial pressure of B can enhance the associative desorption process, reducing the effective reaction probability.

If we look at some of the CVD processes described in the literature, it is easy to see some of these effects. In the case of Si CVD from SiH_4, a reduction in the growth rate was observed with increasing H_2 back pressure. We can interpret these results in terms of the kinetic discussion mentioned above or more generally as an expression of Le Chatelier's principle (Fig. 4.9).

In Fig. 4.10, we show one such example, in this case for the CVD of TiB_2 from the $Ti(BH_4)_4$dme (dme = 1,2-dimethoxyethane) precursor. The conformality of this process increases when an additional partial pressure of 1,2-dimethoxyethane, at the cost of a reduction in the growth rate, as could be expected for a decrease in the reaction probability [13].

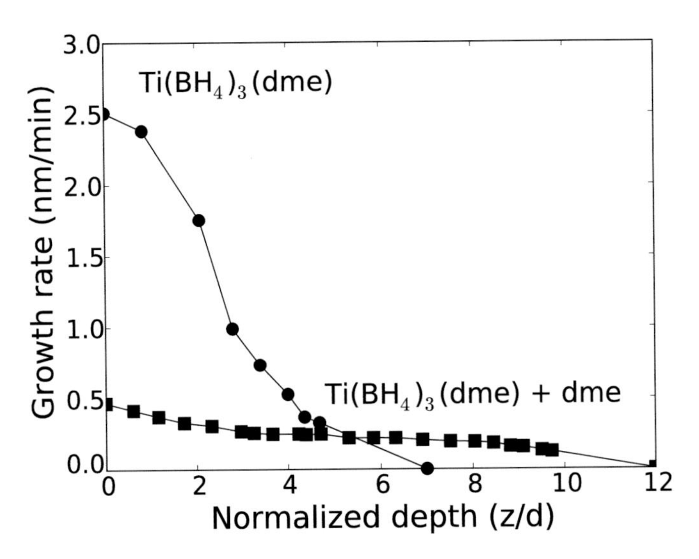

Fig. 4.10 Example of conformality enhancement by growth inhibition: trench growth profile of TiB_2 deposited using $Ti(BH_4)_4$dme as a function of the dme partial pressure. Reproduced from Ref. [13] with permission

4.4 Self-limited Surface Kinetics: Atomic Layer Deposition

In contrast to CVD, where as shown above conformality is strongly dependent on kinetics, the conformal nature of ideal ALD processes makes them ideally suited for the coating of nanostructured materials. The numerous examples of ALD on high surface area materials reported in the literature include high aspect ratio features, anodised alumina [14], nanoparticle and powder beds [15], silica gel [16], nanostructured substrates, colloidal crystals [17], aerogels [18], and the infiltration of polymers [19]. There are several reviews in the literature focused on the fabrication possibilities of ALD with high surface area materials [20, 21].

The key issue in the case of self-limited processes is to understand the kinetics of infiltration as well as the length of the required exposures, and the impact of non-idealities present in many real ALD process. These may include spurious CVD components, the impact of reaction byproducts, and the presence of other channels such as recombination of atomic species in plasma-assisted ALD. Kinetics and non-idealities are the two main reasons behind failure to achieve conformal deposition in ALD. These will be the focus of this section.

4.4.1 *Infiltration Kinetics*

From an experimental point of view, the coating of a high surface area material by ALD requires a modification of the dose times needed to reach saturation when coating flat substrates to ensure that the high surface area material is evenly coated. As shown in Chap. 2, in a conventional ALD process, two reactants A and B are alternatively dosed into the reactor, with a purge time between each pulse to allow for any unreacted precursor molecules or reaction byproducts to be pumped out before the next precursor is introduced. Therefore, a typical ALD process can defined by a tuple t_1, t_2, t_3, t_4, where t_1 and t_3 are the dose times of the ALD precursors and t_3 and t_4 are the corresponding purge times.

Strictly speaking, though, these dose times are just a shorthand way of referring to exposures: the key parameter is not time itself, but the total number of moles that are introduced in the system. This value is proportional to the product of precursor pressure times exposure time. However, the control of the total exposures is typically done by controlling the exposure times rather than by consciously modifying the precursor pressure.

As shown in Chap. 2, control of dose and purge times is crucial to ensure that saturation is reached everywhere and that there is no overlap between the two precursors. The reasons why dose and purge times need to be increased when coating high surface area materials are essentially three: first, the high surface area increases the total number reactive sites. Therefore, longer doses are simply needed to ensure that enough precursor molecules are inserted into the reactor. Second, the transport of species inside a high surface area material requires time. This means that we need to allow for enough time both for the precursor to diffuse through the nanostructured materials, and for the reaction byproducts and the remaining unreacted molecules to leave the reactor during purge times. Finally, the presence of high surface area materials can affect the dynamics of precursor transport at a reactor scale. A high surface area material can act as an sponge exacerbating precursor consumption and reducing the exposures at certain points of the surface [21].

4.4.1.1 Models for ALD Infiltration Kinetics

As mentioned in Chap. 2, in a self-limited process the key variable controlling the surface kinetics is the surface coverage of the precursor Θ. As a surface is exposed to a gas, it reacts with the available surface sites until it reaches saturation, $\Theta \to 1$.

If we focus on the infiltration kinetics of a single precursor, the evolution of the surface coverage can be modeled by the following equation:

$$\frac{d\Theta(\mathbf{x})}{dt} = f(\Theta, \phi(\mathbf{x})) \tag{4.72}$$

The simplest case is the first order irreversible Langmuir kinetic model described in Chap. 2, by which the evolution of the surface coverage is given by:

$$\frac{d\Theta}{dt} = s_0 \frac{1}{4} \bar{v} n \beta_0 (1 - \Theta) \tag{4.73}$$

In this model, the effective sticking probability β is proportional to the fraction of available sites on the surface, so that $\beta = \beta_0(1 - \Theta)$. β_0 that can be interpreted as the bare reaction probability, that is, the probability that an incident molecule irreversibly reacts with a pristine surface. Finally, s_0 represents the average area of a single surface site, so that $1/s_0$ represents the number of surface sites per unit area, which in turn can be related with the growth per cycle through the density of the film. The irreversible first order Langmuir kinetics is by far the most common surface kinetics model used in the literature to model ALD processes, and most of the modeling work on the infiltration in high surface area or nanostructured materials have used this model.

A key difference between ALD and CVD is that we now have three different characteristic times: the time required for species to diffuse into the material, the saturation time at each point of that surface, and the dose time required to fully saturate the complete feature.

Most of the transport models in the literature assume that the saturation time is much larger than the time it takes for the flux inside the feature to equilibrate. The local flux can then be calculated using any of the approaches described in Chap. 3. This was referred to in Ref. [22] as the *frozen surface approximation*. Examples in the literature include approaches based on ballistic transport, such as the work of Kim et al., where they applied their model to the growth of TiO_2 [23, 24], Monte Carlo simulations [25], and diffusion based models [22]. In contrast, a transport model based on a random walk approximation developed to model ALD inside porous anodised alumina model transport and the change in coverage simultaneously [14]. In Refs. [22, 26] a discussion on the conditions under which transport and the change in surface coverage can be decoupled is presented in the context of a diffusion-based model. This analysis is summarized in Sect. 4.4.1.4. However, more research is needed to clarify the validity of this approximation and to establish a connection between the random-walk model of Elam et al. and the transport models presented in Chap. 3.

4.4.1.2 Impact of Reaction Probability on Coating Profiles

The reaction probability β_0 has a strong impact on the coverage profiles inside high surface area materials. One of the first systematic studies on the impact of the reaction probability on the coating of high surface area materials by ALD was carried out by Deendoven et al. [25].

When the reaction probability is high, the coating profiles evolve as a step function, with the saturation front propagating deeper into the film with increasing exposures. Conversely, if the sticking probability is low enough, the surface coverage gradients are small, and saturation is reached evenly inside the material. This is shown in Fig. 4.11, where the profiles are calculated for two cases representative of the high reaction probability and the low reaction probability regimes.

As in the CVD case, the coating of high surface area materials by ALD can be also treated using a continuum model based on the diffusion equation. In this section we

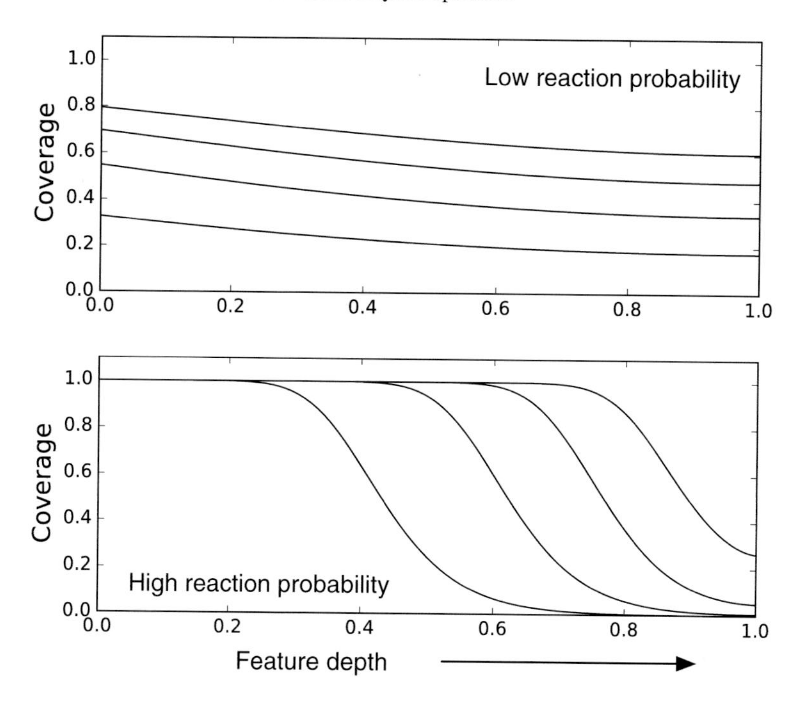

Fig. 4.11 Impact of reaction probability in coverage profiles: as the reaction probability increases growth profile transition to essentially flat to step-profiles

will follow the approach used in Refs. [22, 27], except for a change in nomenclature to emphasize the similarities with the treatment carried out for the constant reaction probability case in Sect. 4.2.

To model ALD processes in high surface area materials, two equations are needed: the first one describes the transport of the precursor within the high surface area material, while the second treats the evolution of the surface coverage at every point:

$$D\frac{\partial^2 n}{\partial z^2} = \bar{s}\beta_0 \frac{1}{4}\bar{v}\,(1-\Theta)\,n \tag{4.74}$$

$$\frac{d\Theta(z,t)}{dt} = s_0\beta_0 \frac{1}{4}\bar{v}\,(1-\Theta)\,n \tag{4.75}$$

Just as in Sect. 4.2, we can transform these two equations into non-dimensional analogs by considering a high surface area material of thickness L and a characteristic value of the precursor density n_0. The resulting equations are:

$$\frac{\partial^2 x}{\partial \xi^2} = h_T^2\,(1-\Theta)\,x \tag{4.76}$$

$$\frac{d\Theta}{ds} = h_T^2\,(1-\Theta)\,x \tag{4.77}$$

where:

$$x = n/n_0 \tag{4.78}$$

$$\xi = z/L \tag{4.79}$$

$$s = t\frac{\gamma D}{L^2} \tag{4.80}$$

$$\gamma = \frac{s_0 n_0}{\bar{s}} \tag{4.81}$$

And h_T is the Thiele modulus defined in Eq. 4.9:

$$h_T^2 = \frac{1\bar{s}\bar{v}\beta_0}{4}\frac{L^2}{D} \tag{4.82}$$

From Eqs. 4.76 and 4.77 it is clear that the kinetics of infiltration in ALD depends solely on the Thiele modulus. This parameter controls how surface coverage evolves towards saturation inside the high surface area material: if $h_T \gg 1$, the saturation profiles approximate a step function, with the saturation front propagating deeper into the feature with increasing exposures. In contrast, if $h_T \ll 1$, coverage profiles are homogeneous and saturation is reached everywhere inside the feature approximately at the same time. Based on the results presented in Sect. 4.2, it can be argued that only when $h_T \gg 1$ we take advantage of the self-limited nature of ALD, since for $h_T \ll 1$ a non-self limited process would also be able to achieve good conformality.

As in the constant reaction probability case, if we particularize the results for a long circular pore, the Thiele modulus reduces to:

$$h_T = \sqrt{3\beta}(\mathrm{AR}) \tag{4.83}$$

Therefore, large reaction probabilities and high aspect ratio favor an infiltration kinetics driven by a saturation front that propagates deeper into the high surface area material, as demonstrated by kinetic Monte Carlo simulations [25].

The second important lesson that we can extract from the non-dimensional equations is that instead of the time t we now have a non-dimensional time s, which is related to the real time through a proportionality constant dependent on a parameter γ. This parameter represents the number of gas molecules per surface site: and was referred to as the *excess number* in Ref. [27]. This proportionality constant will reappear when analyzing the exposure times in the next section.

4.4.1.3 Exposure Times to Coat High Surface Area Materials

In an ideal ALD process, selecting the right exposure and purge times is the key to ensuring that a high surface area or nanostructured material is fully infiltrated in a self-limited way. In this section we are going to focus on the exposure or dose times in an ALD cycle.

The best known estimate is a simple expression found by Gordon et al. for the dose times required to saturate a ciruclar pore of an aspect ratio AR [28]. The main assumption used by the authors was that the growth profile propagates down the circular pore as a step function. They then applied the expression for the transmission probability of particles moving through a circular pore to determine the flux of particles as a function of depth. This is in turn used to determine the velocity at which this front propagates, assuming that all the particle reaching the saturation front react with the surface. The resulting expression for the saturation time:

$$pt = \frac{\sqrt{2\pi M k_B T}}{s_0}\left[1 + \frac{19}{4}\mathrm{AR} + \frac{3}{2}(\mathrm{AR})^2\right] \tag{4.84}$$

Neither in the derivation of Eq. 4.84 nor in Eq. 4.84 itself there is any mention to the surface kinetics: the increase in dose time come from the fact that: 1) as the aspect ratio increases there are naturally more surface sites that need to be covered, and 2) as the aspect ratio increases, the fraction of molecules getting to the bottom of the pores decreases, as shown in Chap. 3. In order to compensate for the decreasing transmission probability, exposures need to be increased even further.

The main assumption of Eq. 4.84 is that the growth profile propagates down the circular pore as a step function. As we have previously seen, this corresponds to the limiting case of a high reaction probability and high aspect ratio.

Yanguas-Gil and Elam derived a generalized expression for an arbitrary value of the sticking probability and for a general nanostructured material [22, 27]. Their analysis, based on a diffusion model for the precursor transport, considered the two limiting cases of the Thiele modulus, referred to as α in their work: the limit $h_T \gg 1$ corresponds to the step-function propagation considered by Gordon et al. However, when $h_T \ll 1$, the saturation is reached evenly inside the feature with almost no precursor depletion: the time to coat the feature is then dominated by the saturation time t_0 given by the solution of Eq. 4.73. This time is independent of the aspect ratio or surface area of the material, and is given by Eq. 2.24:

$$t_0 = \frac{4k_B T}{s_0 p \bar{v} \beta_0}|\log(1 - c_0)| \tag{4.85}$$

In Ref. [22], a comparison between the numerical solution of Eqs. 4.76 and 4.77 and the simple addition of the two asymptotic solutions t_d and t_0:

$$t = t_d + t_0 \tag{4.86}$$

showed an excellent agreement for all values of h_T. Consequently, a simple way of generalizing Eq. 4.84 is to add the saturation time obtained for the same precursor pressure on a flat substrate.

Since Eqs. 4.76 and 4.77 are solved by an arbitrary diffusion constant, it is also possible to generalize Eq. 4.84 to an arbitrary material. The resulting expression is simply:

$$t = t_0 + \frac{1}{\gamma}\frac{L^2}{D} \tag{4.87}$$

where γ is the excess number defined in Eq. 4.81, and the diffusivity D will vary depending on the microstructure of the material, as described in Chap. 3.

4.4.1.4 Frozen Surface Approximation

Almost all the approaches to model the coating of high surface area materials by ALD assume that the equilibration time of the transport inside the high surface area material is much faster than the characteristic saturation time. However, other models, such as that of Elam et al., treated the transport and the evolution of surface coverage simultaneously [14]. In this section we summarize the analysis carried out in Ref. [22] to understand in the context of a continuum model the validity of this assumption.

The starting point is a time-dependent version of Eqs. 4.74 and 4.75:

$$\frac{\partial n}{\partial t} - D\frac{\partial^2 n}{\partial z^2} = -\bar{s}\beta_0\frac{1}{4}\bar{v}\,(1-\Theta)\,n \tag{4.88}$$

$$\frac{d\Theta(z,t)}{dt} = s_0\beta_0\frac{1}{4}\bar{v}\,(1-\Theta)\,n \tag{4.89}$$

If we now normalize this equation in the same way as Eqs. 4.74 and 4.75, we obtain:

$$\gamma\frac{\partial x}{\partial s} - \frac{\partial^2 x}{\partial \xi^2} = -h_T^2\,(1-\Theta)\,x \tag{4.90}$$

$$\frac{d\Theta}{ds} = h_T^2\,(1-\Theta)\,x \tag{4.91}$$

The time derivative in Eq. 4.90 is proportional to the excess number γ, defined as the number of molecules in the gas phase inside the high surface area material per surface site. The argument used in Ref. [22] was that, whenever $\gamma \ll 1$, the time derivative term can be neglected, and the frozen surface approximation holds.

If we particularize the expression of γ for a circular pore, we have that:

$$\gamma = \frac{s_0 n_0}{\bar{s}} = \frac{s_0 d p_0}{4 k_B T} \tag{4.92}$$

For a pressure of 1 Torr, a surface site area $s_0 = 10^{-19}$ m^2, and a temperature of 473 K, the value of γ is roughly:

$$\gamma \approx 500 \times d \tag{4.93}$$

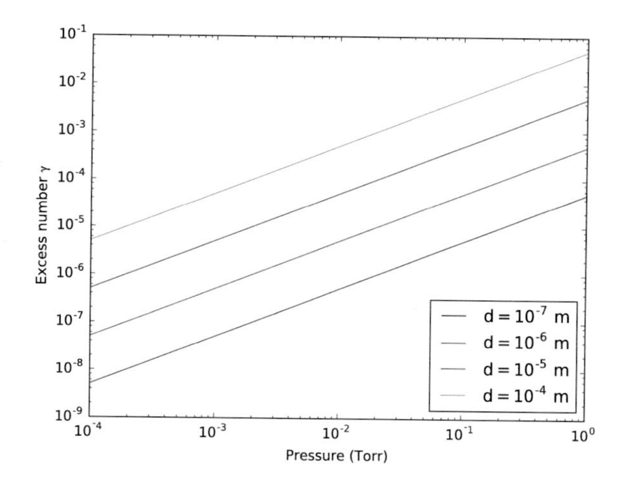

Fig. 4.12 Excess number γ as a function of precursor pressure for different pore sizes. Whenever $\gamma \ll 1$ the transport and surface evolution times can be decoupled

where d is the pore diameter in meters. It is clear that as long as the pores are micron-sized or below, $\gamma \ll 1$, and the frozen surface approximation would hold. Values of γ as a function of precursor pressure for different pore diameters are shown in Fig. 4.12.

It is important to note that there are aspects of the reactive-transport process that are not captured in this model. For instance, as shown in Sect. 3.3.5, the presence of reversible surface adsorption/desorption can slow down the transport of molecules. As a further example, Gobbert et al. carried out detailed simulations based on the Boltzmann equation of the transients of adsorption and desorption of molecules in shallow features, finding that, compared with the transport times, the characteristic times for adsorption/desorption have a much larger impact in the overall dynamics of the system [29].

4.4.2 Non-ideal ALD Processes

In addition to the need of controlling exposure times correctly, a second important factor complicating the coating of high surface area materials in ALD is the presence of non-ideal surface kinetics. These include soft-saturating behaviors, in which the ALD process has fast and a slow surface kinetics components, systems with a small CVD component, the self-inhibiting effect of some ligands, such as the alkyl alcohols in alkoxide precursors, which can block the precursor adsorption and lead to gradients in the reactor even in presence of saturation, and the presence of secondary surface recombination pathways that do not contribute to the ALD process.

Compared to the ideal first order Langmuir kinetic model, only a few cases have been studied from a fundamental point of view. Yanguas-Gil and Elam very briefly considered the case of soft-saturating systems by considering two separate reaction pathways at the surface: a fast pathway characterized by a high reaction probability, and a secondary channel that was at least one order of magnitude slower [26].

Probably the best studied case is the impact of surface recombination, which naturally appears in plasma-enhanced ALD processes. This general case was treated by Knoops et al., who identified the presence of three different regimes: [30] in addition to the transport-limited and the reaction-limited regimes common to thermal ALD processes, the coating of high aspect ratio materials can also take place in a *recombination-limited regime*, in which the time to reach saturation is condition by the ability to reach high exposures of the gas phase species deep inside high aspect ratio features. The key parameter in this case is the recombination probability of the atomic species. This can range from greater than 0.1 for many metals, to lower than 10^{-4} for some oxides such as silica [31, 32]. Note that some of these values can be pressure-dependent.

4.5 Summary

In this chapter we have introduced the fundamentals aspects of thin film growth in nanostructured materials under different conditions: from a CVD perspective, we have focused on understanding the impact that surface kinetics have on the conformality of a process, and how the dependence of growth rate with pressure can be used to derive a criterion that CVD processes must meet in order to conformally coat high surface area reactors. These simplified models can also be used to rationalize approaches to improve the conformality and even achieve superconformal growth. In the case of ALD, we have explored the impact of surface kinetics on the dynamics of infiltration inside high surface area materials and the determination of exposure times required in order to achieve saturation everywhere inside the nanostructured material. In both CVD and ALD the Thiele modulus emerges as the critical parameter controlling the reactive transport of species inside the high surface area material.

References

1. I.A. Blech, Thin Solid Films **6**, 113 (1970)
2. J.B. Bindell, T.C. Tisone, Thin Solid Films **23**, 31 (1974)
3. T.C. Tisone, J.B. Bindell, J. Vac. Sci. Technol. **11**, 72 (1974)
4. T.S. Cale, G.B. Raupp, T.H. Gandy, J. Appl. Phys. **68**, 3645 (1990)
5. F. Meseguer, A. Blanco, H. Miguez, F. Garcia-Santamaria, M. Ibisate, C. Lopez, Colloids Surf. A Physicochem. Eng. Aspects **202**, 281 (2002)
6. H. Miguez, E. Chomski, F. Garcia-Santamaria, M. Ibisate, S. John, C. Lopez, F. Meseguer, J.P. Mondia, G.A. Ozin, O. Toader, H.M. van Driel, Adv. Mater. **13**, 1634 (2001)
7. K.A. Arpin, M.D. Losego, A.N. Cloud, H.L. Ning, J. Mallek, N.P. Sergeant, L.X. Zhu, Z.F. Yu, B. Kalanyan, G.N. Parsons, G.S. Girolami, J.R. Abelson, S.H. Fan, P.V. Braun, Nat. Commun. **4**, 2630 (2013)
8. G.B. Raupp, T.S. Cale, Chem. Mater. **1**, 207 (1989)

9. A. Yanguas-Gil, Y. Yang, N. Kumar, J.R. Abelson, J. Vac. Sci. Technol. A **27**, 1235 (2009)
10. V. Ramachandran, M.F. Brady, A.R. Smith, R.M. Feenstra, D.W. Greve, J. Electr. Mater. **27**(4), 308 (1998)
11. W.B. Wang, J.R. Abelson, J. Appl. Phys. **116**(19), 194508 (2014)
12. W.J.B. Wang, N.N. Chang, T.A. Codding, G.S. Girolami, J.R. Abelson, J. Vac. Sci. Technol. A **32**(5), 051512 (2014)
13. A. Yanguas-Gil, N. Kumar, Y. Yang, J.R. Abelson, J. Vac. Sci. Technol. A **27**(5), 1244 (2009)
14. J.W. Elam, D. Routkevitch, P.P. Mardilovich, S.M. George, Chem. Mater. **15**(18), 3507 (2003)
15. S. Haukka, E.L. Lakomaa, O. Jylha, J. Vilhunen, S. Hornytzkyj, Langmuir **9**(12), 3497 (1993)
16. J.W. Elam, J.A. Libera, T.H. Huynh, H. Feng, M.J. Pellin, J. Phys. Chem. C **114**, 17286 (2010)
17. A. Rugge, J.S. Becker, R.G. Gordon, S.H. Tolbert, Nano Lett. **3**, 1293 (2003)
18. J.W. Elam, J.A. Libera, M.J. Pellin, A.V. Zinovev, J.P. Greene, J.A. Nolen, Appl. Phys. Lett. **89**(5), 053124 (2006)
19. Q. Peng, Y.C. Tseng, S.B. Darling, J.W. Elam, Adv. Mater. **22**(45), 5129 (2010)
20. M. Knez, K. Niesch, L. Niinisto, Adv. Mater. **19**(21), 3425 (2007)
21. A. Yanguas-Gil, J.A. Libera, J.W. Elam, ECS Trans. **64**(9), 63 (2014)
22. A. Yanguas-Gil, J.W. Elam, Chem. Vapor Depos. **18**(1–3), 46 (2012)
23. J.Y. Kim, J.H. Ahn, S.W. Kang, J.H. Kim, J. Appl. Phys. **101**, 073502 (2007)
24. J.Y. Kim, J.H. Kim, J.H. Ahn, P.K. Park, S.W. Kang, J. Electrochem. Soc. **154**, H1008 (2007)
25. J. Dendooven, D. Deduytsche, J. Musschoot, R.L. Vanmeirhaeghe, C. Detavernier, J. Electrochem. Soc. **156**, P63 (2009)
26. A. Yanguas-Gil, J.W. Elam, Theor. Chem. Acc. **133**(4), 1465 (2014)
27. A. Yanguas-Gil, J.W. Elam, ECS Trans. **41**(2), 169 (2011)
28. R.G. Gordon, D. Hausmann, E. Kim, J. Shepard, Chem. Vapor Depos. **9**, 73 (2003)
29. M.K. Gobbert, S.G. Webster, T.S. Cale, J. Electrochem. Soc. **149**(8), G461 (2002)
30. H.C.M. Knoops, E. Langereis, M.C.M. van de Sanden, W.M.M. Kessels, J. Electrochem. Soc. **157**(12), G241 (2010)
31. B.J. Wood, H. Wise, J. Phys. Chem. **65**(11), 1976 (1961)
32. Y.C. Kim, M. Boudart, Langmuir **7**(12), 2999 (1991)

Chapter 5
Advanced Concepts

This chapter focuses on two fundamental aspects of the thin film growth of nanostructured materials. The first one is the evolution of the shape of nanostructured materials as a consequence of the growth process. Having the ability to model how this shape evolves as a function of time allows us to predict the structure of the resulting material and fine-tune the growth conditions to ensure that growth remains fully conformal. This is particularly useful whenever full infiltration is desired. Typical examples include trench and via filling in semiconductor manufacturing, the fabrication of photonic materials though gas-phase infiltration in sacrificial scaffolds, and the fabrication of composite materials through chemical vapor infiltration. Luckily, there are a number of approaches that have been developed to model this process and that could be incorporated to the nanofabrication toolbox. The first part of this chapter introduces these methods.

The second part of this chapter deals with the impact of high surface area materials at the reactor scale. The presence of nanostructured materials can greatly impact the transport of reactive species at a reactor scale, in some cases acting as a perfectly reacting surface, causing the appearance of localized areas where the precursor is strongly depleted. Section 5.2 in this chapter introduces the modeling approaches used to bridge the coating of nanostructured materials with the reactive transport of species at a reactor scale.

5.1 Predicting Shape Evolution

As the thickness of the deposited material increases, the the inner surfaces of a high surface area or nanostructured material will evolve until, if the film becomes thick enough, the material is either completely filled or pinch-off or clogging occurs within the nanostructure.

Chapter 4 dealt with the problem of understanding the impact that surface kinetics had on the conformality of the process, but it neglected the problem of surface evolution. In order to properly model the growth of a film of a thickness comparable to the average pore size or the dimensions of a nanostructured feature, the evolution of

A. Yanguas-Gil, *Growth and Transport in Nanostructured Materials*,
SpringerBriefs in Materials, DOI 10.1007/978-3-319-24672-7_5

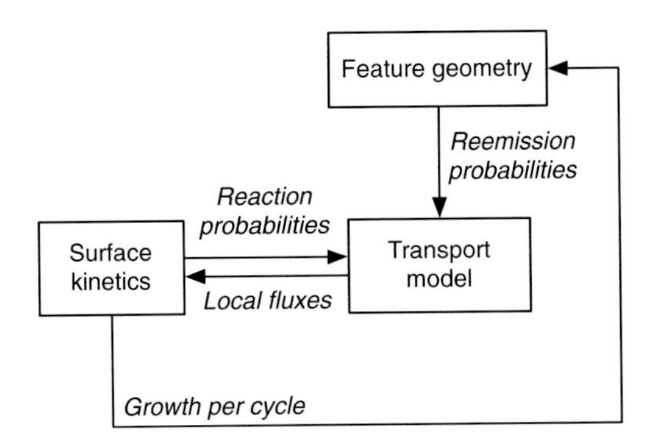

Fig. 5.1 Algorithm to model feature evolution under reactive transport. Reproduced from Ref. [20] with permission

the inner surface, and consequently a reassesment of the conformality of the process, must be considered as the thickness of the film increases.

The resulting algorithm is typically similar to that shown in Fig. 5.1. An inner loop in the model computes the transport and local fluxes of the relevant species coupled to a model of the surface kinetics. The resulting growth rates are then used to calculate the evolution of the inner walls and the overall shape. The process is then repeated until the desired thickness is obtained.

The focus of this section is therefore to review the different approaches that can be used to model surface evolution. These methods can be classified in two broad groups: Lagrangian and Eulerian approaches. In addition to these types of models, there are specific modes that have been created to address specific geometries, such as the constriction of long pores. These will be introduced in Sect. 5.1.7.

5.1.1 Lagrangian Methods

Lagrangian methods track the evolution of the interface in either a 2D or a 3D space. If we define Γ as our surface, it can be represented in parametric form as $\mathbf{x}(s_1, s_2; t)$, where s_1 and s_2 are two parameters that define the surface in space for every given time t. In the case of translational symmetry, for instance in the case of an infinitely long trench, it is possible to reduce the problem to a 2D representation, in which the surface is represented by the function $\mathbf{x}(s, t)$, where s is a real variable and t is time. This 2D representation has been commonly used for the simulation of growth and etching of long trenches or fins, and also features that are axially symmetric, such as circular pores or vias. Within this 2D representation, three natural parameters are the outward normal vector to the surface, $\widehat{\mathbf{n}}$, the tangent vector to the curve $\widehat{\mathbf{t}}$, and the local curvature κ of the surface. In the more general 3D case, the surface representation is characterized by two linearly independent tangent vectors $\widehat{\mathbf{t}_1}$ and $\widehat{\mathbf{t}_2}$.

A natural approach to simulate surface growth is to model the evolution of $\mathbf{x}(s, t)$ as a function of time given a local growth rate $F(s, t)$. In the simplest case in which surface relaxation processes are not important, the change in surface topography will be given by the equation:

$$\frac{\partial \mathbf{x}}{\partial t} = F(s,t)\widehat{\mathbf{n}} \tag{5.1}$$

Models that focus on numerically solving Eq. 5.1 or 3D generalizations are typically referred to as Lagrangian models. These include the string, ray tracing algorithms and the method of characteristics, introduced in Sects. 5.1.3 and 5.1.4 below.

5.1.2 Eulerian Methods

Eulerian approaches do not treat the surface Γ as an independent entity, but it is defined in terms of one or more functions $c(\mathbf{x})$ defined in the whole space. The surface is defined in terms of a contour of this surface, so that:

$$\Gamma \equiv \{\mathbf{x}|c(\mathbf{x}) = c_0\} \tag{5.2}$$

These methods are referred to as Eulerian methods, and the cell and level set methods introduced in Sects. 5.1.5 and 5.1.6 are two of the most common examples.

5.1.3 String Methods

String methods tackle the problem of interface evolution by discretizing Γ using either a string of markers in 2D or a polygonal mesh in the case of 3D surfaces. These approaches have long been used to model both growth and etching during semiconductor processing, with their origin stretching back to the work of Jewett et al., who applied this approach to the problem of photoresist development [1].

If we focus on the 2D case, which is the most commonly used in the literature, string methods discretize a curve $\mathbf{x}(s,t)$ as a polyline composed of straight segments delimited by a series of marker points or nodes $\{\mathbf{x}_i\}$, so that a segment s_i is delimited by the pair of nodes $(\mathbf{x}_i, \mathbf{x}_{i+1})$, as shown in Fig. 5.2.

There are many works that have used string methods to model surface evolution. Many of these works approach the problem of interface evolution in the context of surface etching rather than thin film growth, but the underlying algorithms are the same in both cases. Here we focus on two of the most common approaches: the angle bisector method and the segment advancement method. The method of characteristics is treated separately in Sect. 5.1.4.

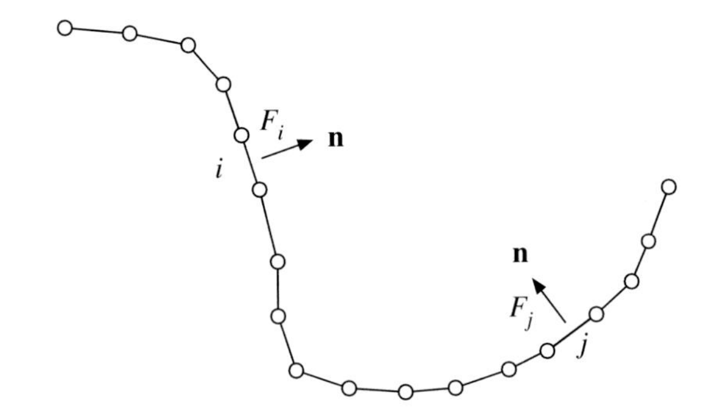

Fig. 5.2 Lagrangian approach to interface evolution, in which the surface is tracked using a series of markers that evolve as a function of time in a way that is proportional to the local flux of species

5.1.3.1 Angle Bisector Method

In the angle bisector method, the normal at each node $\mathbf{x}_i$ is determined by bisecting the intersection angle between the two adjacent segments sharing $\mathbf{x}_i$, as shown in Fig. 5.3. If F_i and F_{i+1} are the local growth rates at the two segments adjacent to ξ, then the advancement of $\mathbf{x}_i$ is determined as follows:

$$\mathbf{x}_i(t + \Delta t) = \mathbf{x}_i(t) + \frac{F_i + F_{i+1}}{2} \Delta t \, \widehat{\mathbf{n}}_i \tag{5.3}$$

where $\widehat{\mathbf{n}}_i$ is the normal local to $\mathbf{x}_i$.

5.1.3.2 Segment Advancement Method

In the segment advanced method, segments are moved according to the local growth rate, with the position of the nodes being determined by the intersection of the two segments (Fig. 5.3).

Let $\widehat{\mathbf{t}_\mathbf{i}}$ be the unit vector parallel to segment i between the points $\mathbf{x}_i$ and $\mathbf{x}_{i+1}$ and $\widehat{\mathbf{n}_\mathbf{i}}$ the normal vector to the segment. If the growth rate of segments i is given by F_i, after Δt segment i is displaced by an amount $F_i \Delta t$. Consequently, $\mathbf{x}_i(t + \Delta t)$ will be contained within the displaced segment, and given by the following equation:

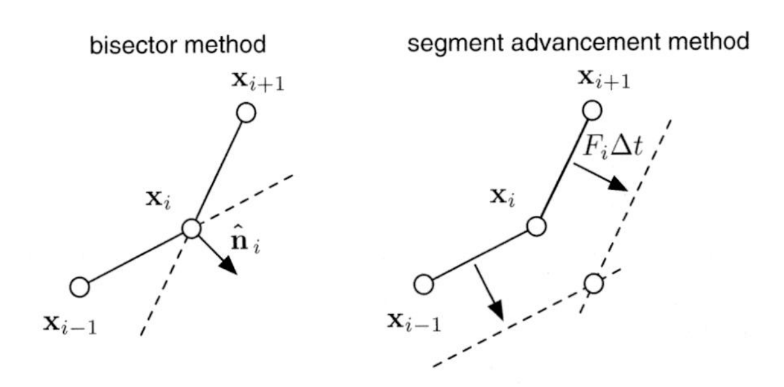

Fig. 5.3 Comparison between the bisector and segment advancement methods used to evolve a surface used by string algorithms

$$\mathbf{x}_i(t + \Delta t) = \mathbf{x}_i(t) + F_i \Delta t\, \widehat{\mathbf{n}}_i + \xi_i \widehat{\mathbf{t}}_i \tag{5.4}$$

where ξ_i is a yet-undetermined scalar.

The same reasoning can be carried out for segment $i + 1$, resulting in the equivalent expression:

$$\mathbf{x}_i(t + \Delta t) = \mathbf{x}_i(t) + F_{i+1} \Delta t\, \widehat{\mathbf{n}}_{i+1} + \xi_{i+1} \widehat{\mathbf{t}}_{i+1} \tag{5.5}$$

The values of ξ_i and ξ_{i+1} can be obtained by equating these two expressions:

$$\mathbf{x}_i + F_i \Delta t\, \widehat{\mathbf{n}}_i + \xi_i \widehat{\mathbf{t}}_\mathbf{i} = \mathbf{x}_i + F_{i+1} \Delta t\, \widehat{\mathbf{n}}_{i+1} + \xi_{i+1} \widehat{\mathbf{t}}_{i+1} \tag{5.6}$$

If we now project this equation into the directions given by the unit vectors $\widehat{\mathbf{t}}_i$ and $\widehat{\mathbf{t}}_{i+1}$, we get the following system of equations for ξ_i and ξ_{i+1}:

$$\xi_i = (\widehat{\mathbf{t}}_i \cdot \widehat{\mathbf{n}}_{i+1}) F_{i+1} \Delta t + (\widehat{\mathbf{t}}_i \cdot \widehat{\mathbf{t}}_{i+1}) \xi_{i+1} \tag{5.7}$$

$$\xi_{i+1} = (\widehat{\mathbf{t}}_{i+1} \cdot \widehat{\mathbf{n}}_i) F_i \Delta t + (\widehat{\mathbf{t}}_{i+1} \cdot \widehat{\mathbf{t}}_i) \xi_i \tag{5.8}$$

While widely used in the literature, string methods present several challenges: first, the length of the segments changes according to the growth, increasing (decreasing) in size in convex (concave) areas of the surface. This is true for both bisector and the segment advancement algorithms. The consequence is that in convex areas we lose resolution as the growth proceeds. Segment advancement methods also preserve the slope of the original segments. Consequently, they are unable to predict the emergence of new faces unless dynamic insertion of new nodes is implemented in the model.

A second problem is the development of loops. In Fig. 5.4 we show an example of loop formation: in concave areas of the surface, growth causes the progressive shortening of the segment length. When the new intersections are calculated based on the local growth rates, this can lead to segments crossing each other. In order to avoid this problem, either small simulation times or the development of delooping algorithms are required in order to maintain a physically correct representation of the growing surface. Another approach involves the dynamic removal of nodes [2].

A final challenge of these techniques is their treatment of discontinuities of the local surface slope, typically referred to as shocks. These discontinuities can develop in areas such as the bottom corners of a trench or via, or can disappear with increasing film thickness, for instance as sharp edges are rounded up during thin film growth.

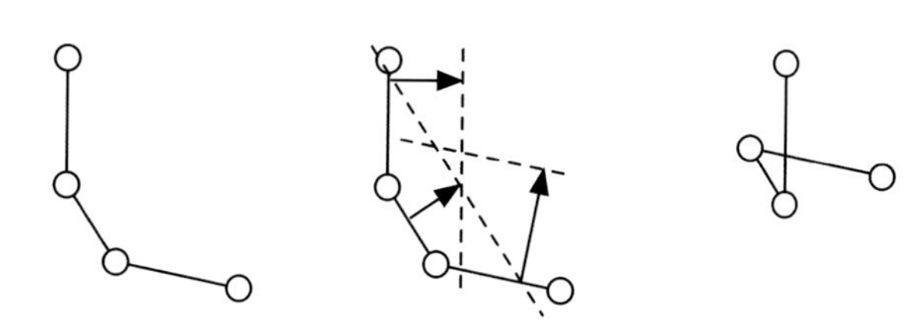

Fig. 5.4 Loop formation in the segment advancement method

Despite the shortcomings mentioned above, string methods have been extensively used in the context of growth and etching both in research codes and commercial TCADs. For instance, the results shown in Fig. 4.3 in Chap. 4 from Cale et al. [2] were obtained using this approach.

5.1.3.3 Extension to 3D Surfaces

While these methods have mostly been applied in the scientific literature for 2D cases, a generalization to 3D is described by Scheckler and Neureuther [3]. The authors used a triangular mesh representation of the surface containing and tracking the evolution of node points, edge segments, and triangular facets. From a computational standpoint, their data structure was composed of three separate lists of triangles, segments, and points, with each triangle containing references to the segments and points forming the facet, each segment containing reference to the points but also the adjacent triangles, and each point list keeping track of the facets they belong to and also of their prior position. Their list was dynamic, with elements being added and deleted according to the evolution of the mesh. One of the challenges of 3D string methods is how to effectively identify and remove unphysical loops that may appear as the surface evolves.

5.1.4 *Method of Characteristics*

The method of characteristics developed by Ross [4, 5] is an alternative approach that takes advantage of our knowledge of the local growth rate to predict the evolution of the surface and the emergence and disappearance of shocks. This can be used to develop efficient strategies to prevent the formation of unphysical loops and therefore it has been used in 2D simulations of topography evolution during etching and growth in semiconductor processing applications. Here we are going to follow the discussion presented in Ref. [6], with some modifications in notation.

5.1.4.1 Surface Evolution Equation

For simplicity, let us focus on a surface that can be locally represented by a function:

$$y = u(x, t) \tag{5.9}$$

While the presence of overhangs and complex structures is directly taken into account using a parametrization such as that shown in Eq. 5.1, at least locally it is always possible to represent the surface using Eq. 5.9 by properly adjusting our coordinate system. The goal is to obtain the equation that models the evolution of the surface for any given growth rate.

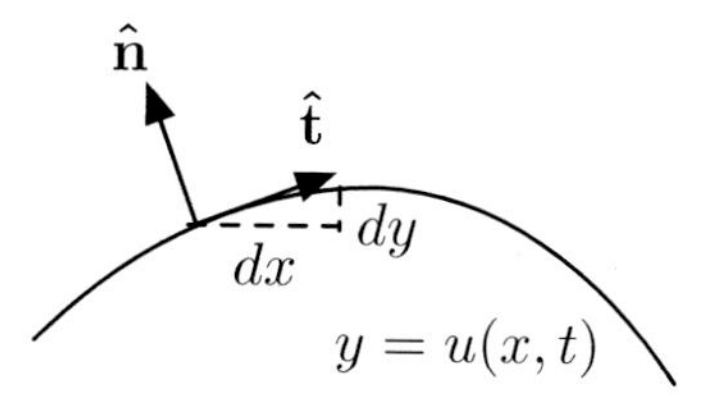

Fig. 5.5 A surface can be locally represented by a function $y = u(x, t)$, which allows us to define the slope, normal and tangent vectors in terms of $p = dy/dx = u_x$

If $\widehat{\mathbf{t}}$ is the tangent vector to the surface, then we can express this vector in terms of Eq. 5.9 as follows:

$$\widehat{\mathbf{t}} = \frac{\mathbf{e}_x + u_x \mathbf{e}_y}{\sqrt{1 + u_x^2}} \tag{5.10}$$

where we have used the shorthand notation:

$$u_x \equiv \frac{\partial u}{\partial x} \tag{5.11}$$

We can then define a normal vector at the same point as:

$$\widehat{\mathbf{n}} = \frac{-u_x \mathbf{e}_x + \mathbf{e}_y}{\sqrt{1 + u_x^2}} \tag{5.12}$$

The physical meaning of u_x is that it represents the local slope p of the surface, since essentially:

$$u_x = \frac{dy}{dx} = p \tag{5.13}$$

as shown in Fig. 5.5.

We can now represent the growth rate or velocity of the interface at a point $\mathbf{x}$ in terms of a vector $\mathbf{F}$ that can be expressed in the most general sense as follows:

$$\mathbf{F} = F\widehat{\mathbf{n}} + F'\widehat{\mathbf{t}} \tag{5.14}$$

The growth rate can have a growth component that is normal to the surface and a second component that is tangential to the surface at every point $\mathbf{x}$. In general, $\mathbf{F}$ will depend on the local position, time, and also, in the most general case, on the slope p at that point of the surface. This vector $\mathbf{F}$ determines how a point $\mathbf{x}$ of the surface will change with time. Consequently we can write the equation:

$$\frac{d\mathbf{x}}{dt} = \mathbf{F} = F\widehat{\mathbf{n}} + F'\widehat{\mathbf{t}} \tag{5.15}$$

We can now use the expressions of $\widehat{\mathbf{t}}$ and $\widehat{\mathbf{n}}$ to get the equations for the evolution of the x and y components of $\mathbf{x}$:

$$\frac{dx}{dt} \equiv \dot{x} = -F\frac{u_x}{\sqrt{1+u_x^2}} + F'\frac{1}{\sqrt{1+u_x^2}} \tag{5.16}$$

$$\frac{dy}{dt} \equiv \dot{y} = F\frac{1}{\sqrt{1+u_x^2}} + F'\frac{u_x}{\sqrt{1+u_x^2}} \tag{5.17}$$

where again we are using the shorthand notation $\dot{x}$ to represent the derivative with respect to time.

On the other hand, we know that $\dot{x}$ and $\dot{y}$ are related by Eq. 5.9, so that:

$$\dot{y} = u_x\dot{x} + u_t \tag{5.18}$$

where we have only calculated the derivative with respect to time and applied the chain rule.

We can now combine these three equations to extract an equation ruling the evolution of the surface, resulting in the expression:

$$u_t = F\sqrt{1+u_x^2} \tag{5.19}$$

Note that only the normal component to the growth rate contributes to surface evolution for an infinite surface.

5.1.4.2 Application of the Method of Characteristics

We can use Eq. 5.19 as a starting point to find a way of propagating the surface that is mathematically correct. To do so, we will apply the method of characteristics, a general technique used to obtain solutions of partial differential equations.

If u is a function of x and t, $u(x,t)$ and we denote $p = u_x$ and $q = u_t$, Eq. 5.19 can be expressed as:

$$q = F\sqrt{1+p^2} \tag{5.20}$$

Equation 5.20 is a particular case of a general non-linear first order partial differential equation, which can be expressed as:

$$F(x,t,u,p,q) = 0 \tag{5.21}$$

The philosophy of the method of characteristics is to identify trajectories $(x(s), y(s))$ so that $u(x(s), t(s))$ are solutions of the PDE along those trajectories. The easiest way of finding such trajectories is to focus first on the evolution of $p(s)$ and $q(s)$ and find those trajectories that simplify the value of p and q along those curves.

Starting with p, if we apply the chain rule we have that:

$$\frac{dp}{ds} = p_x\frac{dx}{ds} + p_t\frac{dt}{ds} \tag{5.22}$$

One of the challenges of this equation is that it depends on the second derivative of u, since $p_x = u_{xx}$ and $p_t = u_{xt}$. To see how we can get rid of these second derivatives, let's differentiate Eq. 5.21 with respect to x. Applying the chain rule, we obtain that:

$$\frac{dF}{dx} = F_x + F_u u_x + F_p p_x + F_q q_x = 0 \tag{5.23}$$

Rearranging this equation and realizing that $q_x = p_t = u_{xt}$, we have that:

$$F_p p_x + F_q p_t = -F_x - F_u u_x = -F_x - F_u p \tag{5.24}$$

By comparing Eqs. 5.22 and 5.24, we see that if we set:

$$\frac{dx}{ds} = F_p \tag{5.25}$$

and

$$\frac{dt}{ds} = F_q \tag{5.26}$$

then we can substitute Eq. 5.24 in Eq. 5.22 and we obtain:

$$\frac{dp}{ds} = -F_x - F_u p \tag{5.27}$$

This, in a nutshell, is the method of the characteristics: we take an equation such as Eq. 5.21 and transform it into a set of ordinary differential equations by choosing trajectories where p and q behave in a simple way.

If we now take Eq. 5.19 and apply the procedure above, the following equations are obtained:

$$\frac{dx}{dt} = -F_p\sqrt{1+p^2} - F\frac{p}{\sqrt{1+p^2}} \tag{5.28}$$

$$\frac{dy}{dt} = -F_p p\sqrt{1+p^2} + F\frac{1}{\sqrt{1+p^2}} \tag{5.29}$$

$$\frac{dp}{dt} = (F_x + pF_u)\sqrt{1+p^2} \tag{5.30}$$

Equations 5.28–5.30 can be simplified further if instead of using p, we employ the angle variable θ defined as $p = \tan\theta$, where θ takes values in the range $-\pi/2 \leq \theta \leq \pi/2$. Then we have that:

$$\frac{dx}{dt} = -F_\theta \cos\theta - F \sin\theta \tag{5.31}$$

$$\frac{dy}{dt} = -F_\theta \sin\theta + F \cos\theta \tag{5.32}$$

$$\frac{d\theta}{dt} = F_x \cos\theta + F_u \sin\theta \tag{5.33}$$

where:

$$F_\theta = \frac{\partial F}{\partial \theta} = \frac{F_p}{\cos^2\theta} \tag{5.34}$$

From here, we can extract some conclusions: first, if the growth rate F does not explicitly depends on θ, the characteristic curve is perpendicular to $y = u(x,t)$. Furthermore, if F is constant, the angle θ is preserved and the characteristics become straight lines.

Equations 5.31–5.33 can be used to evolve the discretized surface in a way that is mathematically correct.

5.1.4.3 Understanding Growth on Sharp Edges

When a curve is discretized as an array of segments, it creates discontinuities in the slope of the curve at every marker $\mathbf{x}$ of the discretized curve. Some of these discontinuities, though, will be real, like the sharp edge of a step or the corner at the bottom of a V-shape trench or via.

We can also use Eq. 5.19 as a starting point to analyze the nature of these discontinuities and determine which of them have physical meaning and which are artifacts of the discretization process.

Let $\mathbf{x}^-$ and $\mathbf{x}^+$ be two points at either side of a slope discontinuity. We will then have that:

$$y^- = u(x^-, t) \tag{5.35}$$

$$y^+ = u(x^+, t) \tag{5.36}$$

Since the curve is continuous at this point, we have that:

$$u(x^-, t) = u(x^+, t) \tag{5.37}$$

Differentiating with respect to time and applying the chain rule, we have that:

$$u_x(x^-, t)\dot{x} + u_t(x^-, t) = u_x(x^+, t)\dot{x} + u_t(x^+, t) \tag{5.38}$$

Let us use a shorthand notation, and express Eq. 5.19 as:

$$u_t = f(x, t, u, p) \tag{5.39}$$

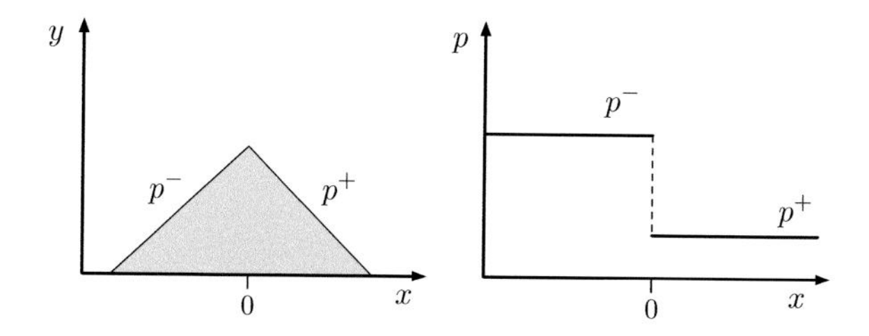

Fig. 5.6 A sharp edge on a surface represents a discontinuity on its slope p. Such discontinuity is made explicit if we represent the slope as a function of x. Using the gas dynamics analogy, this can be viewed as a shock, with the slope playing the same role as pressure

where

$$f = F\sqrt{1 + u_x^2} \tag{5.40}$$

Then, from Eq. 5.38 we obtain the following rule for the evolution of a point at a discontinuity:

$$\dot{x} = \frac{f^+ - f^-}{p^- - p^+} \tag{5.41}$$

This equation rules the evolution of the slope discontinuity with time, that is, the displacement in the x axis of the discontinuity and it is called the jump condition. Interestingly, the same expression appears in gas dynamics in the context of the study of pressure shocks. In that context it is known as the Rankine-Hugoniot condition.

Equation 5.41 dictates how the position of slope discontinuities evolves with time. However, some of these discontinuities may become unstable and disappear as a function of time. This is something that intuitively we know from observing growth around a sharp edge. The coating progressively blunts the edge transforming what originally can be considered a discontinuous front into a rounded continuous curve.

In order to tacke this issue, let's differentiate Eq. 5.19 as a function of x, the resulting equation:

$$u_{tx} = p_t = \partial_x \left(F\sqrt{1 + p^2}\right) \tag{5.42}$$

provides an equation on the evolution of slopes as a function of time. Using our definition of f, Eq. 5.42 reduces to:

$$p_t = \partial_x f(x, t, u, p) \tag{5.43}$$

This equation tracks the evolution of the slope as a function of position and time. Therefore, a region around the sharp edge depicted in Fig. 5.6 would be represented as a discontinuity in the value of p at that position, assumed in Fig. 5.6 to be the origin.

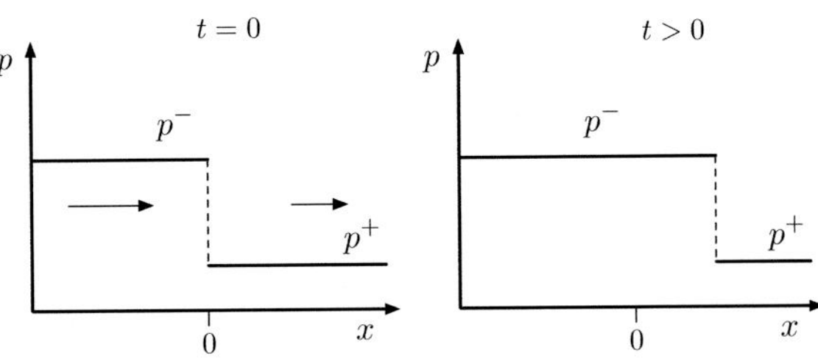

Fig. 5.7 By analogy with gas dynamics, if the velocity at which the slope propagates before the discontinuity is larger than the velocity after the discontinuity, we expect the edge to remain sharp

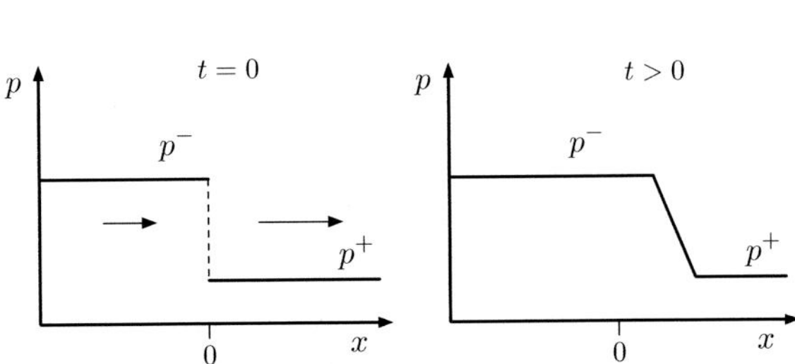

Fig. 5.8 By analogy with gas dynamics, if the velocity at which the slope propagates before the discontinuity is smaller than the velocity after the discontinuity, we expect the edge to become blunt

We can use the analogy with gas dynamics to define a condition that will tell us when this will happen in terms of the growth parameters and the surface topography: this is what it is commonly referred to in the literature as the *entropy condition*.

Let's consider the evolution around the sharp edge shown in Fig. 5.6. We can have two different scenarios: if we think of Fig. 5.6 as a gas, if the area before the shock moves faster than the area after the shock, we expect the shock to propagate as time increases. This is the scenario depicted in Fig. 5.7.

On the other hand, if the area before the shocks moves slower than the area after the shock, we expect the shock to dissipate and instead a smooth transition is developed between the two slopes. This leads to smoothening, as shown schematically in Fig. 5.8

This is the so-called entropy condition: in order for the shock to exist, the following inequality must be fulfilled:

$$\frac{df}{dp}(p^+) > \frac{df}{dp}(p^-) \tag{5.44}$$

Note that the sign is actually the opposite of what it is generally found in the literature. The reason for this is that in Eq. 5.43 the sign of f is the opposite to the convention used in the literature and that obtained for an etching process.

Let's assume the simple case of a growth rate that does not depend on the local slope. Then, we have that:

$$f(p) = F\sqrt{1+p^2} \tag{5.45}$$

And the inequality (5.44) can be expressed as:

$$\frac{p^-}{\sqrt{1+(p^-)^2}} < \frac{p^+}{\sqrt{1+(p^+)^2}} \tag{5.46}$$

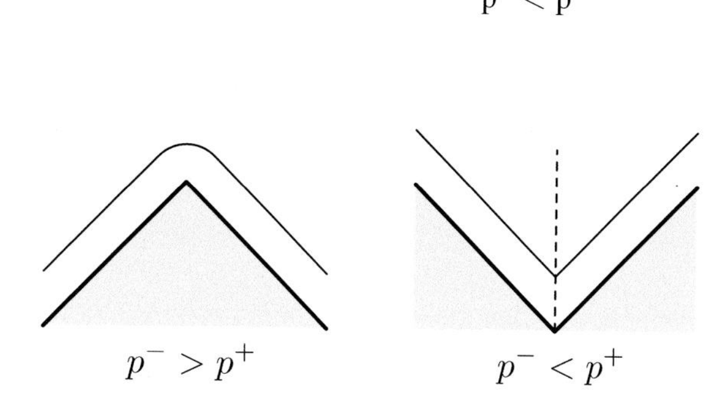

Fig. 5.9 The entropy condition for an isotropic growth rate around a slope discontinuity explains why sharp tips get rounded while V-shaped crevasse maintain their slope discontinuity

which can be simplified even further to:

$$p^- < p^+ \tag{5.47}$$

Equation 5.47 expresses the entropy condition in simple terms, summarized in Fig. 5.9: for discontinuities that extend upwards, like sharp edges, so that the slope before the discontinuity (uphill, and therefore positive) is larger than the slope after the discontinuity (downhill, and therefore negative), no shock can develop (since $p^- > p^+$), and therefore growth results on the blunting of the edge. In contrast, for discontinuities that extend downwards, like sharp recessed or V shaped trenches, the shock will be preserved during growth.

5.1.4.4 Algorithms for Surface Evolution

In addition to contributing to our fundamental understanding of surface evolution during thin film growth, the method of characteristics also allows us to formulate physically correct algorithms to model surface evolution. One such example is the algorithm of Hamaguchi et al. [6]. Their approach is based on treating the union between two segments as a potential shock, using the entropy condition to determine how to treat the discontinuity in the slope introduced by the discretization process. Their algorithm considers three cases:

1. If the entropy condition is satisfied, then Eq. 5.41 is used to determine the evolution of the marker.
2. if the entropy condition is not satisfied, then the segments and decomposed into a series of smaller segments so that the change of slope is below a certain predetermined threshold value $\Delta\theta$.
3. if the change of slope in the segment is below $\Delta\theta$, the points are evolved with time using the characteristic equations derived in the previous section.

By developing a rational approach to surface evolution, algorithms based on the method of characteristics bypass some of the *ad hoc* assumptions made by string

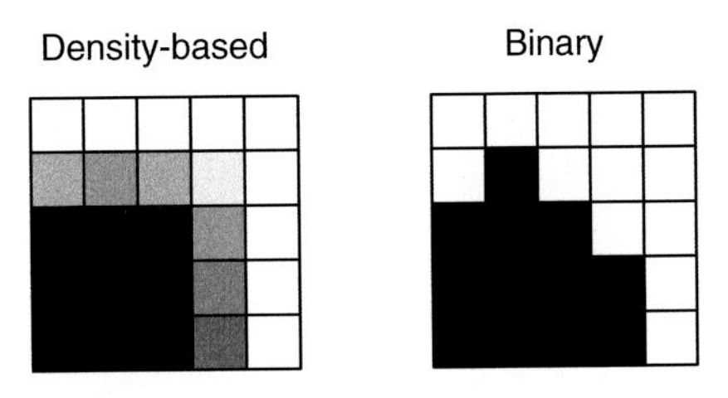

Fig. 5.10 Cell or volume of fluid method

methods, tracking the evolution of the surface in a way that is compatible with the entropy condition.

5.1.5 *Cell or Volume of Fluid Methods*

Cell methods, also known as volume of fluid methods, are one example of an Eulerian approach to surface evolution. Here the space is discretized in a series of elements, and then each cell is assigned a value based on their composition (Fig. 5.10). The approaches in the literature range from allowing multiple materials, to density-based binary materials, or to simpler binary (solid/gas) representations.

One of the challenges of these methods is how to identify where the surface is located and which is its normal, which is of great importance particularly when these methods are to be coupled with ballistic transport models.

One possible approach followed for instance by Fujinaga and Kotani is to consider that the normal direction coincides with the direction of the cell concentration gradient at the surface [7]. In their cell model, for each cell they defined a variable $C(x, y, z)$ defined so that $C = 1$ represents the bulk material, and $C = 0$ the gas phase. The position of the surface is therefore defined as a threshold value $C = 0.5$, with the position of the surface normal being defined by the direction of the concentration gradient ∇C at the surface position.

Then, the evolution of the surface is determined by considering a mass balance equation, by which the increase in $C(x, y, z)$ is given by the local flux to the surface $R(x, y, z)$. For the cases for which the total mass per cell, given by $R(x, y, z)\Delta t$, exceeds the total mass contained within a cell, a spill-over algorithm is introduced that transfers that mass excess to the neighboring cells [7].

A simplification of this approach is to consider each element either empty or full, thereby trivially determining the presence of the interface. This is the approach followed by Coronell and Jensen, who tracked the growth evolution in a trench by discretizing the trench and then using a Monte Carlo method to determine the impingement rates at each point of the surface [8].

5.1.6 *Level Set Methods*

Level set methods were introduced by Osher and Sethian to track the movement of complex interfaces [9]. The central idea of these methods is that the moving front is modeled as a level or contour set of a higher dimensional function. One of the key advantages is that they treat the evolution of kinks and edges in a physical way. However, they involve more complex algorithms, requiring the numerical solution of a hyperbolic partial differential equation.

There are two main approaches: the time-dependent approach is the most general method, valid for arbitrary growth or etch functions and 3D structures. The time-independent approach is simpler to implement, but it is restricted for only pure growth or pure etching conditions.

5.1.6.1 Model Formulation

In this section we are going to follow the approach by Adalsteinsson and Sethian [10].

Let Γ be the set of points in space defining the surface of a nanostructured material or a substrate feature. We can then define the function $\phi(\mathbf{x}, t = 0) = \pm d$, where d is the distance of $\mathbf{x}$ to the surface Γ, taking a positive (negative) value when $\mathbf{x}$ is outside (inside) the material. It is clear then that

$$\phi(\mathbf{x}, t = 0) = 0 \tag{5.48}$$

contains all the points of our interface, that is:

$$\Gamma(t = 0) = \{\mathbf{x}|\phi(\mathbf{x}, t = 0) = 0\} \tag{5.49}$$

The approach of level set methods is to model the evolution of our surface $\Gamma(t)$ with time in terms of the function $\phi(\mathbf{x}, t)$. In order to accomplish this, if $\mathbf{x}(t)$ is a point of the interface at time t, it will have to satisfy the condition:

$$\phi(\mathbf{x}(t), t) = 0 \tag{5.50}$$

We can now differentiate this equation as a function of time. Using the chain rule, we obtain that:

$$\frac{d\phi}{dt} = \frac{\partial\phi}{\partial t} + \nabla\phi \cdot \dot{\mathbf{x}}(t) = 0 \tag{5.51}$$

Here, $\nabla\phi$ represents the gradient of ϕ, which will be normal to the growing surface, and $\dot{\mathbf{x}}(t)$ represent the derivative with time of $\mathbf{x}$.

The question is how we can connect $\dot{\mathbf{x}}(t)$ with the growth rate at the surface: if $F(\mathbf{x})$ represents the growth rate in the normal direction to the surface at the point $\mathbf{x}$, we can see that $F(\mathbf{x})$ is simply the projection of $\dot{\mathbf{x}}(t)$ to the normal of the surface:

$$F(\mathbf{x}) = \dot{\mathbf{x}}(t) \cdot \widehat{\mathbf{n}} \tag{5.52}$$

where $\widehat{\mathbf{n}}$ is the normal vector to the surface at $\mathbf{x}$.

The final step is to realize that we can express $\widehat{\mathbf{n}}$ in terms of the gradient of $\phi(\mathbf{x}, t)$, so that:

$$\widehat{\mathbf{n}} = \frac{\nabla\phi}{|\nabla\phi|} \tag{5.53}$$

Substituting Eq. 5.53 in Eq. 5.51 we obtain:

$$\frac{\partial\phi}{\partial t} + F(\mathbf{x})\,|\nabla\phi| = 0 \tag{5.54}$$

5.1.6.2 Example

We can use a simple example to see the physical meaning of Eq. 5.54. Let us consider the coating of a circular fiber of radius R by a process with a constant growth rate F. Then, Eq. 5.54 reduces simply to:

$$\frac{\partial\phi}{\partial t} + F\left|\frac{\partial\phi}{\partial r}\right| = 0 \tag{5.55}$$

If we consider the function $\phi(r, 0) = \phi_0(r) = r - R$ as our initial condition, then the solution to the partial differential equation is given by:

$$\phi(r, t) = \phi_0(r - Ft) = r - Ft - R \tag{5.56}$$

If we now use the fact that the position of our surface Γ is given by the condition:

$$\Gamma(t) = \{\mathbf{x}|\phi(\mathbf{x}, t) = 0\} \tag{5.57}$$

we obtain that $\Gamma(t)$ is defined by the equation:

$$r = R + Ft \tag{5.58}$$

That is, our surface is a circle that expands radially as a constant velocity given by the growth rate F, just as we would expect.

5.1.6.3 Implementation Challenges

One of the key advantages of the level set method is that Eq. 5.54 can be numerically solved using finite difference schemes. However, care must be taken to use the right approximation to the gradient term in Eq. 5.54, one that propagates the surface accurately without introducing numerical artifacts in the solution. One such scheme was derived by Osher and Sethian and it is provided here as an example. If D_{ijk}^{+x} represents the difference operator,

$$D_{ijk}^{+x} = \frac{\phi_{i+1jk} - \phi_{ijk}}{\Delta x} \tag{5.59}$$

and we assume a growth rate that is everywhere positive, that is, no etching is involved, then then Eq. 5.54 can be discretized as follows:

$$\begin{aligned} \phi_{ijk}^{n+1} = \phi_{ijk}^{n} - \Delta t F_{ijk} \Big(& \max(D_{ijk}^{-x}\phi, 0)^2 + \min(D_{ijk}^{x}\phi, 0)^2 \\ & + \max(D_{ijk}^{-y}\phi, 0)^2 + \min(D_{ijk}^{y}\phi, 0)^2 \\ & + \max(D_{ijk}^{-z}\phi, 0)^2 + \min(D_{ijk}^{z}\phi, 0)^2 \Big)^{1/2} \end{aligned} \tag{5.60}$$

One of the drawbacks of this model is that it requires advancing all the level sets of the function ϕ, not only the $\phi = 0$ level corresponding to the nanomaterial surface. In order to accelerate the calculations, the so-called narrow-band algorithm was developed, which reduced the number of points from $O(N^3)$ to $O(kN^2)$, where k is a finite number of point forming a narrow band surrounding the advancing interface.

5.1.6.4 Time-Independent Formulation and Fast-Marching Algorithms

A faster, simpler approach to the level set method can be obtained whenever growth and etching are not simultaneously present in the system. Under this condition, the evolution of the surface is irreversible: once the surface has reached a point in space that point of space remains coated forever. Under this condition, it is then possible to define a function $T(\mathbf{x})$ containing the time t at which the surface of the nanostructured material reaches that point [11].

If F again represents is the growth rate at every point of space, it is clear that the rate of change of T is related to F through the equation:

$$F\,|\nabla T| = 1 \tag{5.61}$$

This basically says that the time it takes for the surface to evolve from a point $\mathbf{x}$ to a point $\mathbf{x} + d\mathbf{x}$ is simply determined by the growth rate F at $\mathbf{x}$.

One of the advantages of this approach is that it can be solved using a highly efficient algorithm called the fast marching method. The discretization of Eq. 5.61 is done in a slightly different way, using a less-diffusive approximation to the gradient term, resulting in the following equation:

$$\left[\max\left(\max(D_{ij}^{-x}T,0)^2, -\min(D_{ij}^{x}T,0)\right)^2\right.$$
$$\left.\max\left(\max(D_{ij}^{-y}T,0)^2, -\min(D_{ij}^{y}T,0)\right)^2\right] = \frac{1}{F^2} \tag{5.62}$$

subject to the boundary condition:

$$T(\mathbf{x}) = 0 \tag{5.63}$$

for all the points **x** contained within the region of the space occupied by the solid fraction of the nanostructure. The key realization from Eq. 5.62 is that the solution propagates sequentially in time: point not yet coated are affected solely by those neighboring points that have been reached by the surface. This lead Sethian to propose the following iterative algorithm to calculate $T(\mathbf{x})$:

1. First, tag the points in the grid with three different values: boundary value points are tagged as *Known*, and all their nearest neighbors as *Trial*. The rest of points are tagged as *Far*
2. Compute the T values of the points listed as *Trial*
3. Switch the point with the minimum value from *Trial* to *Known*.
4. Tag as *Trial* the nearest neighbors of the selected point.
5. Repeat from 2 until the grid has been filled.

This algorithm provides a extremely efficient way of calculating the evolution of the surface. It is based on the assumption that, since points that are coated remain coated, the solution must propagate upwards from these points to the nearest points of the surface, in a way implementing Huygen's principle recursively.

This algorithm can be generalized to triangular or tetragonal meshes, and details on the procedure can be found in the literature [12].

5.1.6.5 Application to Surface Evolution During Thin Film Growth

Sethian [12] provides several examples of the application of level set methods to model the evolution of surfaces under etching, growth, and ion milling. One of the challenges of this method is the need to recalculate the global view factors as the surface evolves with time. In an approach proposed by Hsiau and coworkers [13], which relies on the polygonal reconstruction of the growth front to recalculate the surface fluxes and therefore determine the local value of the front speed F required by the level set method.

In addition to the examples mentioned above in the context of VLSI, level set methods have also been applied to understand the densification of nanofiber substrates by chemical vapor infiltration. One such example is shown in Fig. 5.11. In this particular example, surface evolution is coupled with a continuum model for the reactive transport of CVI precursors to determine the evolution of a bundle of aligned fibers [14].

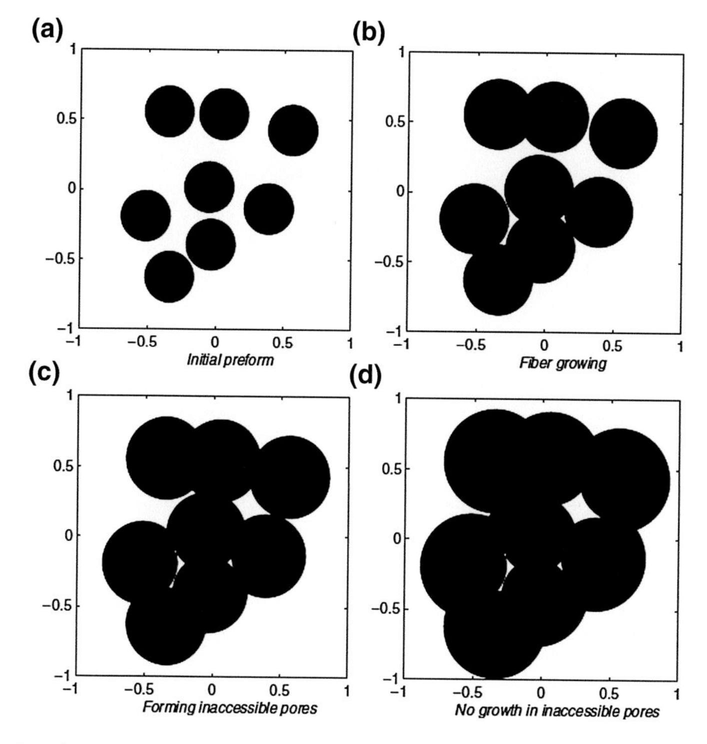

Fig. 5.11 Level set methods allow for the evolution of complex surfaces, such as the fiber bundles represented in this Figure. Reproduced from Ref. [14] with permission

5.1.7 Pore Constriction Models

In prior sections we have presented general approaches to treat the evolution of the inner surface of a nanostructured substrate or high surface area material during thin film growth. In this section we briefly introduce specific models that have been particularized for the reactive transport inside long circular pores. These structures have been used in many areas, and other materials such as anodised aluminum oxide and glass capillary arrays can be represented by these models.

As we showed in Chaps. 3 and 4, by defining a Knudsen diffusivity it is possible to model the reactive transport of gaseous precursors inside high surface area materials using a diffusion equation even when the average diameter of a pore is smaller than the mean free path of the species. This model is also applicable to circular pores, where the Knudsen diffusivity is defined as:

$$D = \frac{1}{3}\bar{v}d \tag{5.64}$$

where d is the diameter of the pore and $\bar{v}$ is the mean thermal velocity, defined as:

$$\bar{v} = \sqrt{\frac{\pi k_B T}{8 M_m}} \tag{5.65}$$

In order to model the filling of high aspect ratio vias, Raupp and Cale considered that, if the spatial scale in which d changes with depth is larger than the via diameter, one can define a local diffusivity $D(z)$ [15]:

$$D(z) = \frac{1}{3}\bar{v}d(z) \tag{5.66}$$

Therefore, the transport equation can be obtained from the balance equation:

$$\frac{\partial\left(A(z)n\right)}{\partial t} = \frac{\partial}{\partial z}\left(A(z)D(z)\frac{\partial n}{\partial z}\right) - rP(z) \tag{5.67}$$

where $A(z)$ is the cross sectional area, $P(z)$ is the perimeter, and r is the reaction rate per unit surface area of the heterogeneous process.

The evolution with time of the cross sectional A is given by a separate equation:

$$\frac{\partial A}{\partial t} = -\frac{rP(z)}{\rho} \tag{5.68}$$

where ρ is the density of the material in atoms per unit volume.

A similar approach was used by Elam et al. for their transport model based on the random walk approximation [16]. They discretized a long tube into an array of segments with constant diameters d_i. Then they treated the diffusion in each of these elements separately, which in their work was modeled as a random walk where particles were allowed to displace a distance of $\pm d_i$ inside of each array.

The main approximation here is that the variations in diameter take place at a length scale that is larger than the diameter of the pore. A necessary condition, but not sufficient, is that the aspect ratio of the pore is large. The pore diameter d_i can be then reduced by a factor proportional to the growth rate in that segment.

5.2 Bridging Feature and Reactor Scales

The presence of high surface area materials can have a strong impact on the transport of reactive species at a reactor scale, particularly when transport takes place under collisional flow. Experimentally, the presence of high surface area materials increases precursor consumption both in CVD and ALD processes, and it can lead to higher inhomogeneities across the substrate.

From a simulation standpoint, different approaches have been developed to incorporate the impact of high surface area materials at a reactor scale. This requires bridging the reactive transport taking place at two very different length scales, and typically two different regimes: if for many CVD processes reactor-scale transport can be modeled using fluid dynamics models, in many cases growth inside nanostructured materials takes place in the Knudsen regime, where the mean free path of molecules is much larger than the characteristic dimension of the nanostructured material, such as the trench width or the average pore size. Understanding how to bridge these two length scales is the focus of this section.

Earlier attempts in the literature directly simulated transport at the reactor and the feature scale sequentially, effectively establishing a one-way coupling between both length scales. Concentrations predicted by the reactor scale model were used as input in the feature scale model to compute the growth inside features. Obviously, these approaches could not account for the impact of additional surface area in the overall reactive transport of species at the reactor scale.

The two-way coupling between the two length scales was introduced in the context of semiconductor processing in the mid 1990s. In this section we summarize the two main approaches developed in the literature: the effective reactivity approach introduced by Jensen and co-workers and the mesoscopic scale approach from Gobbert et al. [17, 18].

5.2.1 *Effective Reactivity Approach*

The effective reaction probability approach, introduced by Rodgers and Jensen [17], establishes a coupling between the reactor and the feature scale at the boundary condition. Instead of modeling the surface topography with high fidelity, a straight control surface is defined and the impact of the high surface area is taken into account through an effective reaction probability for the reactive species and by setting mass flows for reaction byproducts (Fig. 5.12).

If the precursors are diluted in a carrier gas, as it is typically done in many CVD and ALD processes, the reactive species constitute only a small fraction of the total mass flowing through the reactor, and consequently the impact of heterogeneous processes on the overall flow dynamics is small.

Under these conditions, only the transport of reactive species is affected by the presence of a nanostructured substrate. At a reactor scale, the relevant equation is therefore the mass balance equation of the reactive species, which assuming a low

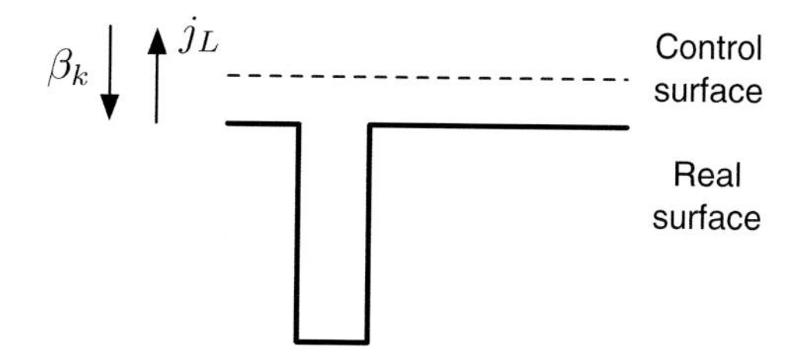

Fig. 5.12 Two way coupling between the reactor and microscopic scales through a control surface

Knudsen number it is given by:

$$\frac{\partial}{\partial t}(\rho Y_k) + \nabla \cdot (\rho \mathbf{u} Y_k) = -\nabla \mathbf{j_k} + G_k \tag{5.69}$$

Here, Y_k represents the mass fraction of species k in the gas phase, $\mathbf{j_k}$ is the diffusion flux, and G_k is the source term in the gas phase given by the local balance of homogeneous reactions.

Under the diluted approximation, the flow velocity $\mathbf{u}$ and the temperature are determined by the momentum and energy balance equations, while the density ρ is found from the ideal gas law.

The connection with the feature scale model takes place through an effective loss term due to the net flow of species at the control surface. If $\tilde{A}$ represents the real surface profile and A the control area, a relationship between the local flux of species and the loss rate at the surface is given as follows:

$$\int j_k dA = \int j_k d\tilde{A} = \int r_k d\tilde{A} \tag{5.70}$$

The flux of molecules per surface area j_k can be expressed in terms of the flux of incident species ϕ_k and an effective reaction probability β_{eff} so that:

$$j_k = \beta_{\text{eff}} \phi_k \tag{5.71}$$

Rodgers and Jensen further decompose β_{eff} in terms of two different factors: if p_{rh} is the probability that a particle incorporates on a homogeneous surface and the ratio ε_k between the actual and the reference incorporation probabilities:

$$\beta_{\text{eff}} = \varepsilon_k p_{\text{rh}} \tag{5.72}$$

For a process characterized by a constant reaction probability that is independent of the local surface flux, ε_k represents a geometric factor that takes into account the increase of surface area modulated by the transport resistance inside the high surface area material.

If the transport inside the high surface area takes place at large Knudsen number, the effective probability can be directly determined using one of the different approaches used presented in Chap. 3, including matrix-base approaches to ballistic transport, Monte Carlo simulations or Markov chains [19, 20].

5.2.2 *Mesoscale Model Approach*

The mesoscale model approach provides an additional level of refinement to the coupling between the reactor and feature scales, and its original goal was to model

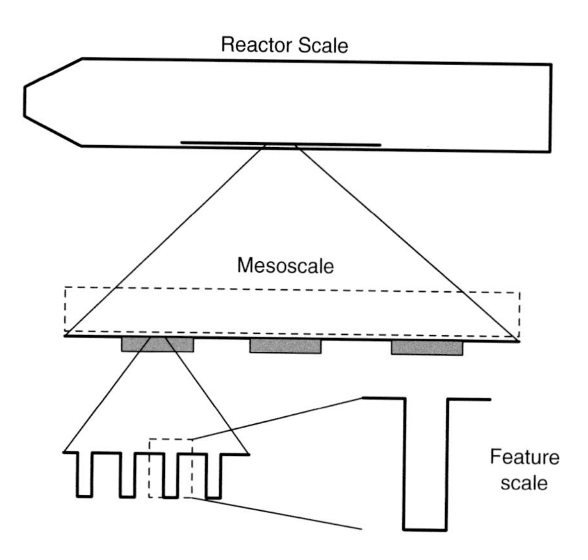

Fig. 5.13 Mesoscale model approach to couple reactor and feature scales for thin film growth simulations

surfaces in which features are aggregated in clusters that are more or less arbitrarily distributed on a wafer. The purpose of the mesoscale model is to connect the reactor with the cluster scales and obtain more detailed information of the local particle fluxes at a cluster scale [18].

The simulation domain of the mesoscale model comprises the gas phase region just above one or more clusters of features. Such an area is represented in Fig. 5.13. The assumption of a low enough Knudsen number coupled with the no-slip velocity condition leads to a simplified equation of precursor transport that is driven by a simple reactor-diffusion equation:

$$\frac{\partial}{\partial t}(\rho Y_k) = -\nabla \dot{\mathbf{j}_\mathbf{k}} + G_k \tag{5.73}$$

The model is coupled with the reactor scale model through the gas-phase boundary condition, whereas the connection with the feature scale model is established through the surface boundary condition. Rather than to follow accurately the surface topography, the mesoscale model uses a flat surface boundary condition at the surface, and instead introduces the effect of nanostructured features through an extra precursor consumption due to the higher surface area.

In the MSM all equivalent features within a cluster are assumed to be coated in an identical way, allowing differentiation at the cluster level. In contrast, when the reactor scale and feature scales are directly coupled, the spatial resolution of feature coating is determined by the discretization of the simulation domain at the reactor scale.

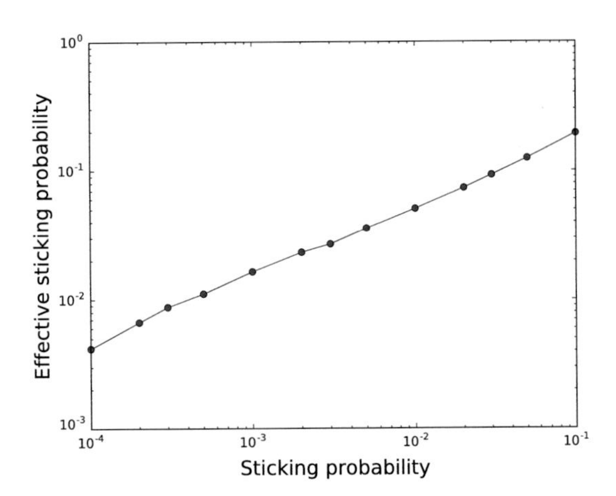

Fig. 5.14 Effective sticking probability for a 12 layer opal for different values of the sticking probability of the underlying surface kinetics

5.2.3 *Effective Reaction Probability of High Surface Area Materials*

The concept of effective reaction probability has a central place in these multiscale approaches, allowing an efficient decoupling between reactor and feature length scales. The question then is how to determine this effective probability and, in the case of ALD, its evolution with coverage.

The main approach mentioned by in the early literature is the use of Monte Carlo simulations [19]. If we use the inverse opal structure shown in Fig. 4.4 in Chap. 4 as an example, by tracking the trajectories of many particles we can determine the probability that the particle reacts *anywhere* within the high surface area material. This is shown in Fig. 5.14 for a reactive system characterized by a constant reaction probability as a function of the sticking probability β_0. Note how the effective reaction probability can be one order of magnitude higher than the process sticking probability β_0. Other approaches to calculate this effective reaction probability include Markov chain models of ballistic transport [20].

5.3 Summary

In this chapter we have focused on two key aspects of thin film growth inside high surface area or nanostructured material: approaches to track surface evolution as the film thickness becomes of the same order as the pore sizes, and the connection between models of reactive transport of the porous material and the reactor scale.

References

1. R.E. Jewett, P.I. Hagouel, A.R. Neureuther, T. Vanduzer, Polym. Eng. Sci. **17**(6), 381 (1977)
2. T.S. Cale, G.B. Raupp, T.H. Gandy, J. Appl. Phys. **68**, 3645 (1990)
3. E.W. Scheckler, A.R. Neureuther, IEEE Trans. Comput.-Aided Des. Integr. Circuits Syst. **13**(2), 219 (1994)
4. D.S. Ross, J. Electrochem. Soc. **135**(5), 1260 (1988)
5. D.S. Ross, J. Electrochem. Soc. **135**(5), 1235 (1988)
6. S. Hamaguchi, M. Dalvie, R.T. Farouki, S. Sethuraman, J. Appl. Phys. **74**(8), 5172 (1993)
7. M. Fujinaga, N. Kotani, I.E.E.E. Trans, Electron Devices **44**(2), 226 (1997)
8. D.G. Coronell, K.F. Jensen, J. Electrochem. Soc. **141**, 2545 (1994)
9. S. Osher, J.A. Sethian, J. Comput. Phys. **79**(1), 12 (1988)
10. D. Adalsteinsson, J.A. Sethian, J. Comput. Phys. **120**(1), 128 (1995)
11. J.A. Sethian, Proc. Natl. Acad. Sci. **93**(4), 1591 (1996)
12. J.A. Sethian, *Level set methods and fast marching algorithms* (Cambridge University Press, Cambridge, 1999)
13. Z.K. Hsiau, E.C. Kan, J.P. McVittie, R.W. Dutton, I.E.E.E. Trans, Electron Devices **44**, 1375 (1997)
14. S. Jin, X.L. Wang, J. Comput. Phys. **179**(2), 557 (2002)
15. G.B. Raupp, T.S. Cale, Chem. Mater. **1**, 207 (1989)
16. J.W. Elam, D. Routkevitch, P.P. Mardilovich, S.M. George, Chem. Mater. **15**(18), 3507 (2003)
17. S.T. Rodgers, K.F. Jensen, J. Appl. Phys. **83**(1), 524 (1998)
18. M.K. Gobbert, T.P. Merchant, L.J. Borucki, T.S. Cale, J. Electrochem. Soc. **144**(11), 3945 (1997)
19. K.F. Jensen, S.T. Rodgers, R. Venkataramani, Curr. Opin. Solid State Mater. Sci. **3**(6), 562 (1998)
20. A. Yanguas-Gil, J.W. Elam, Theor. Chem. Acc. **133**(4), 1465 (2014)

Index

A. Yanguas-Gil, *Growth and Transport in Nanostructured Materials*,
SpringerBriefs in Materials, DOI 10.1007/978-3-319-24672-7

Printed in the United States
By Bookmasters